International Conference on Material Engineering and Manufacturing (ICMEM 2017)

IOP Conference Series: Materials Science and Engineering
Volume 265

Chengdu, China
9 – 11 October 2017

ISBN: 978-1-5108-5264-8
ISSN: 1757-8981

Printed from e-media with permission by:

Curran Associates, Inc.
57 Morehouse Lane
Red Hook, NY 12571

Some format issues inherent in the e-media version may also appear in this print version.

Printed by Curran Associates, Inc. (2017)

For permission requests, please contact the Institute of Physics
at the address below.

Institute of Physics
Dirac House, Temple Back
Bristol BS1 6BE UK

Phone: 44 1 17 929 7481
Fax: 44 1 17 920 0979

techtracking@iop.org

Additional copies of this publication are available from:

Curran Associates, Inc.
57 Morehouse Lane
Red Hook, NY 12571 USA
Phone: 845-758-0400
Fax: 845-758-2633
Email: curran@proceedings.com
Web: www.proceedings.com

Table of contents

Volume 265

2017 International Conference on Material Engineering and Manufacturing (ICMEM 2017)

9–11 October 2017, Chengdu, China

Accepted papers received: 30 October 2017
Published online: 29 November 2017

Preface

Polymers and Composites

Materials Physics

Biomaterials

Materials Processing and Surface Treatment

Mechanical Manufacturing and Automation

Inorganic Nonmetallic Materials

Metal Corrosion and Protection

IOP Conf. Series: Materials Science and Engineering **265** (2017) 011001 doi:10.1088/1757-899X/265/1/011001

PREFACE

It is our great pleasure to introduce you the proceedings of 2017 International Conference on Material Engineering and Manufacturing (ICMEM 2017) held in Chengdu, China during October 9-11, 2017. ICMEM 2017 is dedicated to issues related to material engineering and manufacturing.

One of the objectives of the conference is to establish platforms for collaborative research projects in this field, and to find potential opportunities for international cooperation.

The conference program included keynote, oral, and poster presentations from scholars working in the areas of materials science and engineering. It covered recent trends and progress made in the field of material engineering and manufacturing. Professor from China was invited to deliver keynote speech regarding the latest information in their respective areas of expertise.

These proceedings present a selection from papers submitted to the conference by universities, research institutes, and industries. All the papers were subject to peer-review by conference committee members and international reviewers. The papers were selected based on their quality and their relevance to the conference. The volume presents recent advances in the field of material engineering and manufacturing as well as various related areas, including Materials Science and Engineering, Manufacturing Systems and Equipment, Automation, Control and Information Technology, and Industrial Automation and Process Control, among others.

We would like to express our gratitude to all the members of the conference committee. We would also like to thank the reviewers, who spared their valuable time, for their advice. It has certainly helped improve the quality, accuracy, and relevance of each paper selected for the conference program and for publication. We also wish to thank all the authors who have contributed to this conference, as well as the organizing committee, reviewers, speakers, chairpersons, sponsors, and all the conference participants for their support for ICMEM 2017.

Prof. Xiaohong Zhu, Sichuan University, China
October 20, 2017

IOP Conf. Series: Materials Science and Engineering **265** (2017) 011002 doi:10.1088/1757-899X/265/1/011002

CONFERENCE COMMITTEE CHAIRS

Prof. Xiaohong Zhu, Sichuan University, China
Prof. Dr. WEN Qiye, University of Electronic Science and Technology of China (UESTC), China

CONFERENCE LOCAL CHAIR

Prof. Dr. TANG Xiaoli, University of Electronic Science and Technology of China (UESTC), China

PROGRAM COMMITTEE CHAIRS

Prof. Sofian M. Kanan, American University of Sharjah (AUS), U.A.E
Prof. Shamsuddin B.Sulaiman, Universiti Putra Malaysia, Malaysia
Prof. Arnold C.M.Yang, National Tsing Hua University, Taiwan

TECHNICAL COMMITTEES CHAIRS

Prof. Liqun Zhu, School of Materials Science and Engineering, Beihang University, China
Prof. Sheng-Long Lee, National Central University, Taiwan
Prof. Chiming Huang, Chinese Culture University, Taiwan
Asst. Prof. Bashir Ahmmad, Yamagata University, Japan
Prof. Hern Kim, Myongji University, Korea
Prof. Ir Dr Mariatti Jaafar, Universiti Sains Malaysia, Malaysia
Prof. HJ. Patthi Bin Hussain, Institute of Technology Petronas Sdn. Bhd, Universiti Teknologi Petronas, Malaysia
Prof. Ch. Sanjay, GITAM University Hyderabad, India
Assoc. Prof. DR. KAUSHIK KUMAR, Birla Institute of Technology, India
Dr. Kusno Kamil, Universitas Muslim Indonesia, Indonesia

IOP Conf. Series: Materials Science and Engineering **265** (2017) 011003 doi:10.1088/1757-899X/265/1/011003

Peer review statement

All papers published in this volume of *IOP Conference Series: Materials Science and Engineering* have been peer reviewed through processes administered by the proceedings Editors. Reviews were conducted by expert referees to the professional and scientific standards expected of a proceedings journal published by IOP Publishing.

IOP Conf. Series: Materials Science and Engineering **265** (2017) 012001 doi:10.1088/1757-899X/265/1/012001

Hyperelasticity of Rubbery Materials with Chains' volume, Nonaffine Deformation, and Topological Constraint

D Hao[1,3], L Zhang[1], J Yu[2]*, J Y Shi[1] and D Y Mao[1]

1. High Speed Aerodynamics Institute, China Aerodynamics Research and Development Center, Mianyang, 621000, People's Republic of China
2. Computational Aerodynamics Institute, China Aerodynamics Research and Development Center, Mianyang, 621000, People's Republic of China
3. College of Aerospace Science and Engineering, National University of Defense Technology, Changsha, 410073, People's Republic of China

yujinghd@hotmail.com

Abstract. Rubbery materials play a significant role in industrial field for the reason of better excellent elasticity. A hyperelastic model is proposed, into which chains' volume, nonaffine deformation, and topological constraint are introduced. After obtained the model, it is naturally transformed to investigate homogeneous deformations for isotropic and incompressible rubbery materials and verified by Treloar's experiments. The results show that the present model has ability to express the hyperelasticity of rubbery materials, and the highly predictive accuracy with only simple computation makes the novel model adequate for engineering applications.

1. Introduction

Rubbery materials are able to undergo large, reversible deformations and they have been widely used in industrial applications [1], such as structural bearing [2][3], tires [4], medical devices [5][6], and so on due to their particular mechanical properties. Rubbery materials exhibit a complicated nonlinear behaviour including viscoelasticity [7], Mullins effect [8][9][10], and hyperelasticity [10] etc.. The mechanical response of rubbery materials is usually investigated by numerical simulations, one of them is the finite element method. However, it is still difficult to get the relevant material constants to make them work. So it should be necessary to establish a reliable hyperelastic model for rubbery materials which could be able to express the complex deformations.

The hyperelasticity of rubbery materials can be described by a strain energy density function (SEDF). Once the SEDF is obtained, the stress-stretch relationship can be deduced according to the elastic theory. Two approaches exist to find such a function: one is statistical mechanics and the other is phenomenological theory [10].

In the last few decades, many phenomenological constitutive models, such as Mooney-Rivlin model [11], Gent model [12], Yeoh model [13], Nunes model [14], and Exp-Ln model [15] etc., were proposed or applied by the authors [16][17]. A detailed review for phenomenological models was given by Beta [18]. Although some of the phenomenological models could predict the mechanical properties of rubbery materials well, the coefficients of them lack of physical meanings which are related to the microstructures of polymers. The non-entropic theory of rubbery elasticity for a flexible chain was also detailedly stated by Drozdov [19]. In addition, based on the statistical theory, some micro-mechanical models were also developed. James and Guth [20] proposed a 3-chain model which assumed that the polymer network could be presented by three chains that were oriented in three principle directions. Then, the 3-chain model was improved by Elias-Zunuga in 2006 [21]. Describing the polymer network by means of using four polymer chains, a 4-chain model was presented in [22]. Based on the analysis of the 3-chain and 4-chain models, Arruda and Boyce [23] proposed an 8-chain model (sketched in figure 1). In the 8-chain model, the eight polymer chains are attached to the eight corners of a cube and a center junction. The 8-chain model shows good ability to characterize the behaviour of rubbery materials and it has been widely employed by many authors to express their materials. In recent years, Lopez-Manchado et al. [24] developed a network model with the tube constraints and Kroon [25][26] improved the 8-chain model by considering the topological constrains to account for the non-affine deformation of a polymer chain.

However, most of the previous models for rubbery elasticity are unable to essentially explain the deformation mechanism. Few of them have taken the chains' volume [27], the non-affine deformation, and the topological constraint into account simultaneously. A major objective of our present work is to develop a realistic hyperelastic model to predict the mechanical response of rubbery materials under simple or complex deformations. Introducing the chains' volume, the non-affine deformation, and the topological constraint into the 8-chain model, a novel hyperelastic model is given. Finally, the new model is verified by Treloar's experiments [28] and shows broad prospects in its engineering applications.

2. Network model

We start by considering a cube cell containing eight polymer chains first (see figure 1), which was proposed by Arruda and Boyce [23]. The side length of the undeformed cube is a_0 , and $\mathbf{e}_1$, $\mathbf{e}_2$, and $\mathbf{e}_3$ are the principal directions of deformation. The eight polymer chains are OA, OB, OC, OD, OE, OF, OG, and OH, respectively. After deformation, point O is assumed still at the center of the cube, and the side lengths of the three principal directions separately become $\lambda_i a_0$ (i=1,2,3).

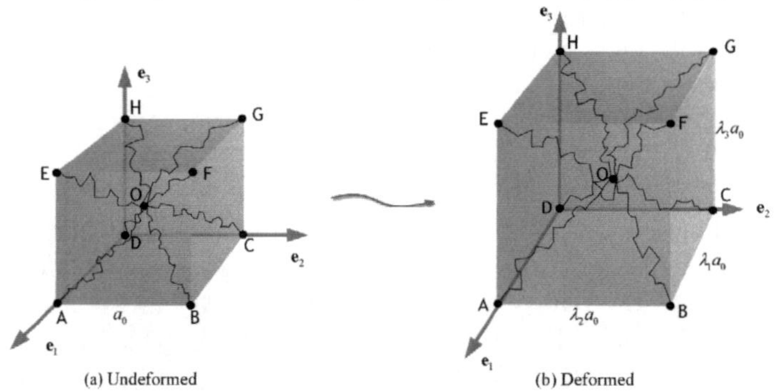

Figure 1. 8-chain model introduced by Arruda and Boyce

A simple chain in figure 1 is taken consisting of N segments of equal length l. The vector between the end points of a chain is $\mathbf{r}$, and the maximum magnitude of $\mathbf{r}$ is $|\mathbf{r}|_{max}=Nl$. It is well-known that the Gaussian distribution theory of rubber elasticity fails to describe the mechanical response of rubbery materials. Thus, the non-Gaussian statistics theory leads to a more realistic molecular distribution function that is valid over the whole range of r up to ultimate or fully extended length. The non-Gaussian probability density function expressed by inverse Langevin is written as follows

$$p_{\text{chain}}(\mathbf{r}) = C_{\text{chain}} \exp\left\{-N\left(\frac{r}{Nl}\beta + \ln\frac{\beta}{\sinh\beta}\right)\right\} \tag{1}$$

where C_{chain} is the normalization constant, $r=|\mathbf{r}|$, and β is defined by Langevin function

$$\coth\beta - 1/\beta = r/Nl \triangleq L(\beta) \tag{2}$$

According to random walk theory, the root-mean-square value of $\mathbf{r}$ for an unstrained chain is r_0=sqrt(N)l. This equation has been employed by many authors to get the chain stretch by $\lambda_c=r/r_0$, such as Elias-Zuniga [21], Kroon [26], and so on. However, the random walk theory ignores the volume and size of the chain, which may be not in accordance with a real polymer chain. In this paper, we take the volume and size of a flexible chain into account. Then the root-mean-square value of $\mathbf{r}$ is assumed as [27]

$$r_0 = N^v l \tag{3}$$

where v describes the effect of the volume and size of the chain. When v=0.5, equation (3) degrades to the result of the random walk theory. If the volume and size of a chain are taken into account, we should easily have that v>0.5, i.e. r_0>sqrt(N)l. Therefore, the stretch of a molecular chain is calculated by $\lambda_c=r/N^v l$. Then, the probability density function expressed by equation (1) is rewritten as

$$p_{\text{chain}}(\lambda_c) = C_{\text{chain}} \exp\left\{-N\left(\frac{\lambda_c}{N^{1-v}}\beta + \ln\frac{\beta}{\sinh\beta}\right)\right\} \tag{4}$$

where $\beta=L^{-1}(\lambda_c/N^{1-v})$.

In addition, the topological network constraint of the polymer network around the cube cell (see figure 1) is also considered in the present model. The constraint probability density function of one side in ei direction is employed as follows in the present work, it is just empirical:

$$p_{\text{con}}(\lambda_i) = C^i_{con} \exp\left(-\frac{h}{\alpha^2\lambda_i^\alpha}\right) \qquad ,i = 1,2,3 \tag{5}$$

where C^i_{con} is the normalization constant, h and α are the parameters used to control the topological network constraint. Here, we let $C^1_{con}=C^2_{con}=C^3_{con}=C^4_{con}=C_{con}$ for simplicity. Using equation (4) and (5), the probability density function of the cube cell (shown in figure 1) could be calculated by

$$p_{\text{cube}} = p_{\text{chain}}{}^8(\lambda_c)\prod_{i=1}^{3} p_{\text{con}}{}^2(\lambda_i) \tag{6}$$

According to Boltzmann's equation, the configuration entropy of the cube cell can be computed by

$$s_{\text{cube}} = k_B \ln p_{\text{cube}} = 8k_B \ln p_{\text{chain}}(\lambda_c) + 2\sum_{i=1}^{3} \ln p_{\text{con}}(\lambda_i) \tag{7}$$

where k_B is Boltzmann's constant. In order to accord with the kinetic theory, we make an assumption that there is no change of the internal energy when the chain deforms. Then, the Helmholtz free energy of the cube cell is given by [35]

$$w_{\text{cube}} = -Ts_{\text{cube}} \tag{8}$$

Submitting equation (4), (5), (6) and (7) into equation (8), we have

$$w_{\text{cube}} = 8Tk_B N\left(\frac{\lambda_c}{N^{1-v}}\beta + \ln\frac{\beta}{\sinh\beta}\right) + 2Tk_B\sum_{i=1}^{3}\left(\frac{h}{\alpha^2\lambda_i^\alpha}\right) + w_0 \tag{9}$$

where T is the absolute temperature and $w_0=8k_B\ln C_{\text{chain}}+6\ln k_B C_{\text{con}}$. If we introduce the chain density n, the strain-energy density function is provided by

$$W = \mu N\left(\frac{\lambda_c}{N^{1-v}}\beta + \ln\frac{\beta}{\sinh\beta}\right) + \frac{2G_{\text{con}}}{\alpha^2}\sum_{i=1}^{3}\left(1/\lambda_i^\alpha\right) + W_0 \triangleq W_{\text{f}} + W_{\text{con}} + W_0 \tag{10}$$

where $\mu=nk_BT$ is the shear modulus and $G_{\text{con}}=\mu h/8$. In equation (10), W_{f} is the strain-energy density function of the free network considering equation (3), and W_{con} is the strain-energy density function caused by the topological constraint for the cube cell. To correlate the microscopic scale deformation to the macroscopic, the homogenization method is put into use here. In this paper, we assume that the stretch of the chain λ_c is non-affine whereas the principal stretch λ_i(i=1,2,3) affine. Thus, the micro-macro transformation of λ_c is given by [20]

$$\lambda_c = K(\lambda_m - 1) + 1 \tag{11}$$

where K is the non-affine coefficient and λ_m is the macroscopic scale of λ_c. For 8-chain model, λ_m is expressed by

$$\lambda_m = \sqrt{\sum_{i=1}^{3}\lambda_i^2/3} \tag{12}$$

When K=1, the transformation of the chain stretch is affine. According to the theory of continuum mechanics, the principal Cauchy stress σ_i of an incompressible material could be calculated from the strain energy density function as [10]

$$\sigma_i = \lambda_i\frac{\partial W}{\partial\lambda_i} - p = \lambda_i\frac{\partial W_{\text{8-chain}}}{\partial\lambda_c}\frac{\partial\lambda_c}{\partial\lambda_m}\frac{\partial\lambda_m}{\partial\lambda_i} + \lambda_i\frac{\partial W_{\text{8-chain}}}{\partial\lambda_i} - p = \mu KN^v\beta\frac{\lambda_i^2}{3\lambda_m} - \frac{2G_{\text{con}}}{\alpha\lambda_i^\alpha} - p \tag{13}$$

in which λ_i are the principal stretches which meet the incompressibility constraint $\lambda_1\lambda_2\lambda_3=1$, and the pressure p is an arbitrary hydrostatic stress that should be determined by the boundary conditions. In order to determine the parameters expediently, the inverse Langevin function can be evaluated by a Pade approximant

$$\beta = \frac{\lambda_c}{N^{1-v}} \frac{3N^{2-2v} - \lambda_c^{\,2}}{N^{2-2v} - \lambda_c^{\,2}} \tag{14}$$

Applying equation (14) to (13), the Cauchy stress σ_i is rewritten as

$$\sigma_i = \mu \frac{K\lambda_c}{\lambda_m} \frac{3N - N^{2v-1}\lambda_c^{\,2}}{3N^{2-2v} - 3\lambda_c^{\,2}} \lambda_i^2 - \frac{2G_{\text{con}}}{\alpha\lambda_i^{\alpha}} - p \tag{15}$$

For the purpose of eliminating p, the following equation is obtained

$$\sigma_i - \sigma_j = \mu \frac{K\lambda_c}{\lambda_m} \frac{3N - N^{2v-1}\lambda_c^{\,2}}{3N^{2-2v} - 3\lambda_c^{\,2}} \left(\lambda_i^2 - \lambda_j^2\right) + \frac{2G_{\text{con}}}{\alpha} \left(1/\lambda_j^{\alpha} - 1/\lambda_i^{\alpha}\right) \tag{16}$$

From equation (16), we can see that the new model degrades to 8-chain model when K=1, v=0.5 and G_{con}=0. This new model owns six parameters, i.e. μ, K, N, v, G_{con}, α. Among them, μ, K and N have definite physical meanings, while v, G_{con} and α are empirical and applied to describing the effect of the volume and size of the chain and the topological constraint empirically, respectively. Before the identification of parameters, it is to be noted that (i) v is larger than 0.5 but smaller than 1, (ii) K varies around 1, and (iii) α is usually close to 1.

3. Parameter identification and model validation

Equation (16) is developed for homogeneous deformations and the nominal stress-stretch relations are provided for simple tension (ST), pure shear (PS), and biaxial tension (BT). This present model is compared with 8-chain model as well as predicting the mechanical response of Treloar's material [28].

3.1 Simple tension

Let us consider a material particle at the position $\mathbf{X}=[X_1\ X_2\ X_3]^{\mathrm{T}}$ in the undeformed reference framework $\mathbf{e}_i$ (i=1,2,3). After deformation, it has a new place $\mathbf{x}=[x_1\ x_2\ x_3]^{\mathrm{T}}$ in the deformed reference framework. Then, a pure homogeneous deformation could be described by

$$\begin{bmatrix} x_1 \\ x_2 \\ x_3 \end{bmatrix} = \begin{bmatrix} \lambda_1 & 0 & 0 \\ 0 & \lambda_2 & 0 \\ 0 & 0 & \lambda_3 \end{bmatrix} \begin{bmatrix} X_1 \\ X_2 \\ X_3 \end{bmatrix} \tag{17}$$

where λ_1, λ_2, and λ_3 are constant principal stretches. Then, the deformation gradient tensor $\mathbf{F}$ is given by

$$\mathbf{F} = \partial\mathbf{x} / \partial\mathbf{X} \tag{18}$$

For an incompressible material, we require that det$\mathbf{F}$= $\lambda_1\lambda_2\lambda_3$=1. In the initial state, $\mathbf{F}=\mathbf{I}$, the unit tensor. Simple tension leads to

$$\lambda_1 = \lambda, \lambda_2 = \lambda_2 = 1/\sqrt{\lambda} \tag{19}$$

$$\sigma_1 = \sigma^{\mathrm{ST}}, \sigma_2 = \sigma_3 = 0 \tag{20}$$

By means of introducing equation (19) and (20) into (16), we obtain the nominal stress-stretch constitutive relationship

$$f^{\mathrm{ST}} = \mu \frac{K\left[K(\lambda_m - 1) + 1\right]}{\lambda_m} \frac{3N - N^{2v-1}\left[K(\lambda_m - 1) + 1\right]^2}{3N^{2-2v} - 3\left[K(\lambda_m - 1) + 1\right]^2} \left(\lambda - \frac{1}{\lambda^2}\right) + \frac{2G_{\text{con}}}{\alpha} \left(\frac{\lambda^{\alpha/2} - 1/\lambda^{\alpha}}{\lambda}\right) = f_f^{\mathrm{ST}} + f_{\text{con}}^{\mathrm{ST}} \tag{21}$$

$$\lambda_m = \sqrt{\left(\lambda^2 + 2/\lambda\right)/3} \tag{22}$$

The expression of $f_{\text{con}}^{\mathrm{ST}}$ in equation (21) was also utilized by Lopez-Manchado et al. [20] to investigate the polymer network of elastomer nanocomposites. Letting v=0.5, K=1, and G_{con}=0 in equation (21), the nominal stress expression of 8-chain model is written as

$$f_{\text{8-chain}}^{\mathrm{ST}} = \mu_{\text{8-chain}} \left(3N_{\text{8-chain}} - \lambda_m^{\,2}\right)\left(\lambda - 1/\lambda^2\right) / \left(3N_{\text{8-chain}} - 3\lambda_m^{\,2}\right) \tag{23}$$

IOP Conf. Series: Materials Science and Engineering **265** (2017) 012001 doi:10.1088/1757-899X/265/1/012001

3.2 Pure shear

A pure shear of homogeneous deformation is described as follows

$$\lambda_1 = \lambda, \lambda_2 = 1, \lambda_3 = \lambda^{-1} \tag{24}$$

$$\sigma_1 = \sigma^{PS}, \sigma_2 = \sigma_2, \sigma_3 = 0 \tag{25}$$

Thus, the nominal stress associated to the stretch is

$$f^{PS} = \mu \frac{K[K(\lambda_m - 1)+1]}{\lambda_m} \frac{3N - N^{2v-1}[K(\lambda_m - 1)+1]^2}{3N^{2-2v} - 3[K(\lambda_m - 1)+1]^2} \left(\lambda - \frac{1}{\lambda^3}\right) + \frac{2G_{con}}{\alpha} \left(\frac{\lambda^{\alpha} - 1/\lambda^{\alpha}}{\lambda}\right) = f_f^{PS} + f_{con}^{PS} \tag{26}$$

$$\lambda_m = \sqrt{\left(\lambda^2 + 1 + 1/\lambda^2\right)/3} \tag{27}$$

Setting v=0.5, K=1, and G_{con}=0 in equation (21), the nominal stress related to 8-chain model is supplied with

$$f_{8\text{-chain}}^{PS} = \mu_{8\text{-chain}} \left(3N_{8\text{-chain}} - \lambda_m^{\ 2}\right)\left(\lambda - 1/\lambda^3\right) / \left(3N_{8\text{-chain}} - 3\lambda_m^{\ 2}\right) \tag{28}$$

3.3 Biaxial tension

A homogeneous biaxial tension is given by

$$\lambda_1 = \lambda_2 = \lambda, \lambda_3 = \lambda^{-2} \tag{29}$$

$$\sigma_1 = \sigma_2 = \sigma^{BT}, \sigma_3 = 0 \tag{30}$$

This leads to the following expression of the nominal stress as

$$f^{BT} = \mu \frac{K[K(\lambda_m - 1)+1]}{\lambda_m} \frac{3N - N^{2v-1}[K(\lambda_m - 1)+1]^2}{3N^{2-2v} - 3[K(\lambda_m - 1)+1]^2} \left(\lambda - \frac{1}{\lambda^5}\right) + \frac{2G_{con}}{\alpha} \left(\frac{\lambda^{2\alpha} - 1/\lambda^{\alpha}}{\lambda}\right) = f_f^{BT} + f_{con}^{BT} \tag{31}$$

$$\lambda_m = \sqrt{\left(2\lambda^2 + 1/\lambda^4\right)/3} \tag{32}$$

If we make v=0.5, K=1, and G_{con}=0 in equation (21), the nominal stress-stretch relationship of 8-chain model could be described by

$$f_{8\text{-chain}}^{BT} = \mu_{8\text{-chain}} \left(3N_{8\text{-chain}} - \lambda_m^{\ 2}\right)\left(\lambda - 1/\lambda^5\right) / \left(3N_{8\text{-chain}} - 3\lambda_m^{\ 2}\right) \tag{33}$$

Hereto, we have got the nominal stress-stretch equations of the new model and 8-chian model. In the following subsection, we utilize the well-known standard experimental data which is published by Treloar [28] for the cases of simple tension, pure shear, and biaxial tension deformations. These results have been widely accepted as representative benchmarks of rubbery materials' response and chosen by many authors to verify their models, such as Arruda and Boyce [23].

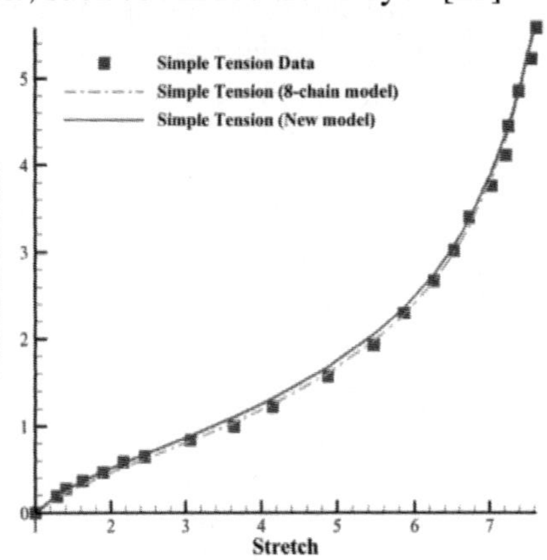

Figure 2. The fitted numerical values of the material parameters in 8-chain model and the new model are identified from the experimental simple tension data by Treloar: $\mu_{8\text{-chain}}$=0.255 MPa, $N_{8\text{-chain}}$=26.12, μ=0.2445 MPa, v=0.502, N=27.9, G_{con}=0.04 MPa, =0.8602, and K=1.023.

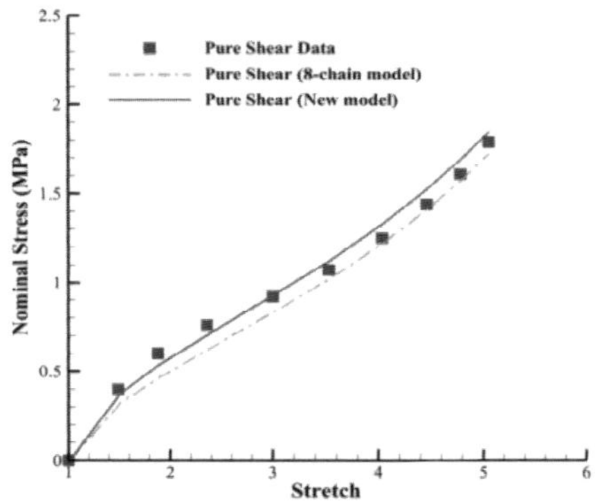

Figure 3. The predicted results of 8-chian model and the new model for pure shear are compared with the experimental pure shear data by Treloar with the parameters identified from the simple tension data alone.

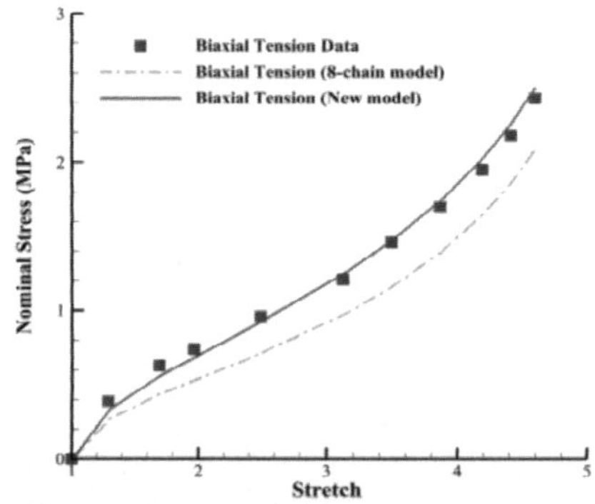

Figure 4. The predicted results of 8-chian model and the new model for biaxial tension are compared with the experimental biaxial tension data by Treloar with the parameters identified from the simple tension data alone.

3.4 Identification of coefficients

The present model has six model parameters and most of them have physical signification which enables to evaluate them from physical measure or bibliographic information. Natheless, all microscopic structural effects are not expressed accurately and numerical values of the model coefficients must be fitted. We use the nonlinear least square method for the identification of the six parameters (μ, K, N, v, G_{con}, α). This method is implemented in the commercial package Matlab^{TM}. Treloar's experimental data [28] is adopted to verify the present model. We choose the simple tension data to fit the parameters in 8-chain model and the new model. The fitted coefficients are illustrated as $\mu_{8\text{-chain}}$=0.255 MPa, $N_{8\text{-chain}}$=26.12, μ=0.2445 MPa, v=0.502, N=27.9, G_{con}=0.04 MPa, α=0.8602, and K=1.023. In figure 2, we compare the fitted results of 8-chain model reported by Arruda and Boyce [23] and the new model with the experimental simple tension data by Treloar [28]. It is shown that each of the two models describes the experimental data of simple tension very well. In addition, v=0.502, K=1.023 and G_{con}=0.04 MPa mean that the volume and size of the chain, the non-affine transformation of the chain stretch, and the topological constraint for the cube cell in Treloar's material maybe exist and should not be ignored in order to obtain an accurate network model. The new model is validated using the pure shear and biaxial tension data. The predicted results of 8-chain model and the new model for pure shear and biaxial tension are displayed in figure 3 and 4, respectively. In figure 3 and 4, we can find out that (i) the new model of equation (16) and 8-chain model both are in well agreement with the experimental data for the pure shear, (ii) 8-chain model underestimates the biaxial tension data which has been discussed in the original paper by Arruda and Boyce [23]. Compared with 8-chain model, the new model provides a better prediction for biaxial tension as well as simple tension and pure shear. The effects of the volume and size of the polymer chain, the non-affine transformation of the chain stretch, and the topological network constraint for the cube cell have improved the predictions of the new model, especially for the biaxial tension. When the volume and size of the polymer chain are considered, the initial end points distance r_0 is larger than sqrt(N)l. The maximum of extensibility of molecular chains is not the same as the one in ideal free networks. The molecular chains reach their extension limit earlier in uniaxial deformation, which was also found out by Bechir, Chevalier, and Idjeri [1].

4. Conclusion

Combining the network model concepts proposed by Arruda and Boyce [23], Lopez-Manchado [24], Doi and Edwards [28] et al., we proposed a new network model to describe the multiaxial elastic response of rubbery materials. In the present model, the volume and size of the polymer chain, the non-affine transformation of the chain stretch, and the topological network constraint in the polymer network are taken into account. The new model uses six parameters (μ, K, N, v, G_{con}, α) that are determined from the only experimental data of simple tension loading. We demonstrate this new model by the experimental data from Treloar [28]. The same model parameters identified from simple tension are applied to predict the response of pure shear and biaxial tension correctly. These parameters are shown to be independent on the state of deformation. The results show that the present model provides an excellent fit to the nominal stress-stretch response of rubbery materials. This new model has the ability to model the elasticity of rubbery materials and could be adapted to engineering applications. The strong predictive abilities with simple computation make it suitable for engineering applications.

References

[1] Bechir H, Chevalier L and Idjeri M 2010 *Int. J. Eng. Sci.* **48** 265–74
[2] White J R and De S D 2001 *Rubber Technologist's Handbook* (Rapra technology limited)
[3] Dick J S 2009 *Rubber Techn.: Compounding Test. Performance* (Hanser Gardner Publications Inc.)
[4] Mark J E, Erman B and Eirich F R 2005 *The Sci. and Techn. of Rubber* (Elsevier academic press)
[5] Loew R and Meier P 2007 *Finite Elem. Anal. Des.* **43** 453–62.
[6] Doyle B J, Corbett T J, Cloonan A J, O'Donnell M R, Walsh M T, Vorp D A and McGloughlin T M 2009 *Medical Eng. Phys.* **31** 1002–12
[7] Shaw M T and MacKnight W J 2005 *Introduction to Polymer Viscoelasticity* (John Wiley @ Sons Inc.)
[8] Machado G, Chagnon G and Favier D 2012 *Mech. Mater.* **50** 70–80
[9] Machado G, Chagnon G and Favier D 2010 *Mech. Mater.* **42** 841–51
[10] Treloar L R G 2005 *The Physics of Rubber Elasticity* (Oxford University Press)
[11] Mooney M 1940 *J. Appl. Phys.* **11(9)** 582–92
[12] Gent A N 1996 *Rubber Chem. Technol.* **69** 59–61
[13] Yeoh O H 1990 *Rubber Chem. Technol.* **63** 792–805.
[14] Nunes L C S 2011 *Mat. Sci. Eng. A-Struct.* **528** 1799–804
[15] Khajehsaeid H, Arghavani J and Naghdabadi R 2013 *Eur. J. Mech. A-Solid.* **38** 144–51
[16] Horgan C O and Murphy J G 2009 *Mech. Mater.* **41** 943–50
[17] Chen Z, Scheffer T and Seibert H 2013 *Mech. Mater.* **64** 111–27
[18] Beda T 2013 *Eur. Polym. J.* (DOI: 10.1016/j.eurpolymj.2013.10.006)
[19] Drozdov A D 2005 *Int. J. Eng. Sci.* **43** 1121–37
[20] James H M, Guth E. 1943 *J. Chem. Phys.* **11** 455–81
[21] Elias-Zuniga A 2006 *Polymer* **47** 907–14
[22] Flory P J, Rehner J J 1943 *J. Chem. Phys.* **11** 512–26
[23] Arruda E M and Boyce M C 1993 *J. Mech. Phys. Solids* **41(2)** 389–412
[24] Lopez-Manchado M A, Valentin J L, Carretero J, Barroso F and Arroyo M 2007 *Eur. Polym. J.* **43** 4143–50
[25] Kroon M 2010 *Mech. Mater.* **42** 873–85
[26] Kroon M 2011 *J. Elasticity* **102** 99–106
[27] Doi M and Edwards S F 1994 *The Theory of Polymer Dynamics* (Oxford University Press)
[28] Treloar L R G 1944 *Trans. Faraday Soc.* **40** 59–70

IOP Conf. Series: Materials Science and Engineering **265** (2017) 012002 doi:10.1088/1757-899X/265/1/012002

Curing kinetics of 4,4'-Methylenebis epoxy and m-Xylylenediamine

Z R Li, X D Li, X Y Guo*
School of Materials and Engineering, Beijing Institute of Technology, Beijing, China

gxy@bit.edu.cn

Abstract. In this paper, the curing kinetics of 4,4'-Methylenebis epoxy resin(TGDDM) and m-Xylylenediamine(m-XDA) was investigated by non-isothermal differential scanning calorimetry(DSC) at various heating rates. Selected non-isothermal methods for analyzing curing kinetics were compared. The activation energy(E) and the correlation coefficient(R) were obtained by different isoconversional methods. The reaction order(n) was obtained by the activation energy in different isoconversional methods for the by Crane equation. The results show that the apparent activation energy are 65.23kJ/mol, 52.20 kJ/mol and 66.10 kJ/mol by using the method of Kissinger, Friedman and F-W-O, the reaction order are 0.911, 0.729 and 0.923 by using the method of Kissinger, Friedman and F-W-O.

1. Introduction

Epoxy are widely used in aviation materials, composite materials and other fields, because it performs very well on the aspects of thermal stability, chemical stability and so on [1,2]. The epoxy resin of 4,4'-Methylenebis perform well in high heat resistance, high strength and high modulus, so it became a hot research [3]. However, the resin usually needs high temperature curing to ensure its performance [4]. High temperature can affect the performance of the resin solidification, consume energy [5].

The performance of material after curing can be affected by kinetic parameters [6]. Therefore, the solidification kinetics of epoxy resins were studied to determine the suitable curing conditions. Although there have been many studies of the curing reaction of epoxy resin at home and abroad, the results are not the same. Therefore, it is necessary to study the curing of specific epoxy resin, curing process for composites forming and optimizing performance.

In this paper, the DSC curves of epoxy resin were measured based on the non- isothermal method, and the reaction kinetic parameters of the epoxy resin system were calculated by Kissinger method, Friedman method and F-W-O method.

2. Experiment

2.1. Raw material.

IOP Conf. Series: Materials Science and Engineering **265** (2017) 012002 doi:10.1088/1757-899X/265/1/012002

4,4'-Methylenebis(AG-80), Shanghai Huayi Resin Co., Ltd, Epoxy value is 0.85; m-Xylylenediamine(m-XDA), Shanghai Macklin Biochemical Co., Ltd.

2.2. Preparation of the sample.
The thermal curing reaction of the AG-80 epoxy resin and m-XDA under flowing N_2 gas was investigated using a DSC analyzer (Netzsch 204 F1). The DSC conditions were as follows: sample mass, 5.000 -8.000 mg; heating rates, 2.5, 5, 10, 15, and 20 K/min; atmosphere, N_2 gas at a flow rate of 20 mL/min; and temperature 30 °C–180 °C.

3. The results and discussion

3.1. DSC curves of Samples.
Figure 1. shows the DSC curves at different heating rate. Table 1. shows the thermal enthalpy ΔH, the initial temperature T_i, the peak temperature T_p and the final temperature T_f.

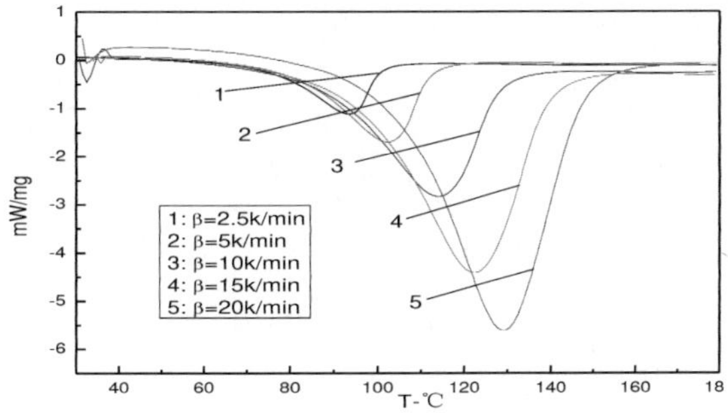

Figure 1. DSC curves of system different heating rates.

Table 1. The respective values of ΔH, T_i, T_p and T_f at different heating rates.

Heating rates (k/min)	ΔH(J/g)	T_i(℃)	T_p(℃)	T_f(℃)
2.5	465.1	76.9	93.7	100.8
5	435.6	82.7	102.4	112.5
10	453.7	89.2	113.9	128.7
15	509.7	96.8	122.4	139.3
20	503.2	104.6	129.1	147.4

3.2. Curing kinetic parameters of method of Kissinger.
Curing kinetic parameters such as apparent activation energy are important to understand the solidification. In general, the Kissinger equations [7,8] can be used to process the DSC data which is at different heating rates. The kinetic parameters can be obtained.

$$\ln\left(\frac{\beta}{T_P^2}\right) = \ln\left(\frac{AR}{E_a}\right) - \frac{E_a}{RT_p} \tag{1}$$

In equation (1), β is the heating rate; T_p is the peak temperature; E_a is the apparent activation energy. E_a and A can be obtained by the relationship between $\ln\beta/T_P^2$ and $1/T_p$. The calculating results show that E_a=65.23kJ/mol, A=3.07×$10^8$$s^{-1}$ and the correlation coefficient R is 0.994.

3.3. Curing kinetic parameters of method of Friedman.
The Friedman method is also used to analyze apparent activation energy. The Friedman equation is shown below.

$$\ln[\frac{\beta d\alpha}{dT}] = \ln Af(\alpha) - \frac{E}{RT} \quad (2)$$

In equation (2), the physical meanings of β, E, T and A are the same as those of equation (1). The differential form of the reaction mechanism function is $f(\alpha)$.

At the different heating rates β, the same conversion rate α is selected, then $f(\alpha)$ is a constant value. So $\ln[\beta d\alpha/dT]$ and 1/T are linear relationship in shown Figure 2, the E and A of can be obtained by the slope and intercept. Using the slope and the intercept of the linear regression line at each conversion rate, the values of E and A of were obtained as shown in Table 2.

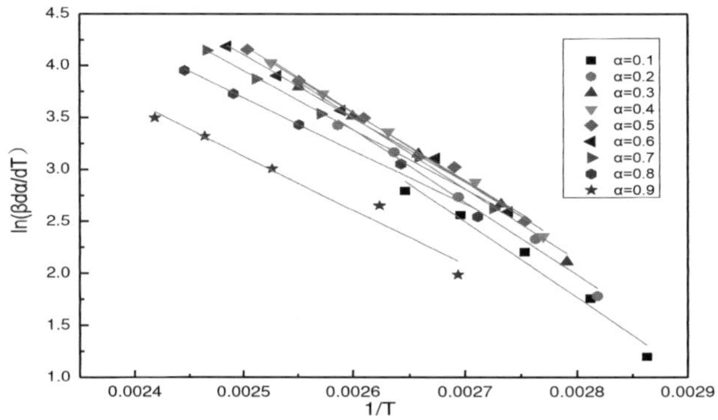

Figure 2. Relationship between $\ln[\beta d\alpha/dT]$ and (1/T).

Table 2. The values of E and A with the method of the Freidman.

α	Intercept	Slope	R	E(kJ/mol)
0.1	22.10	7260.67	0.985	60.37
0.2	21.57	6992.50	0.993	58.14
0.3	21.40	6882.40	0.994	57.22
0.4	20.98	6700.78	0.997	55.71
0.5	20.44	6496.80	0.997	54.01
0.6	19.45	6134.44	0.999	51.00
0.7	18.21	5702.82	0.997	47.41
0.8	16.49	5119.15	0.993	42.56
0.9	16.16	5213.57	0.977	43.35

The E obtained by the Friedman method in the Table 2. were averaged to the obtain the apparent activation energy that E_a=52.20kJ/mol.

3.4. Curing kinetic parameters of method of F-W-O.

The F-W-O [9] method allows to obtain apparent activation energy from a plot of $\lg\beta$ versus 1/T. The Friedman equation is shown below.

$$\lg\beta = \log\left[\frac{AE}{RG(\alpha)}\right] - 2.315 - 0.4567\frac{E}{RT} \quad (3)$$

In equation (3), the physical meanings of β, E, T and A are the same as those of equation (1).

$G(\alpha)$ is the integral form of the reaction mechanism function. Since the same conversion rate is chosen, $G(\alpha)$ is a constant value, so that $\lg\beta$ is linearly related to 1/T.

As with Friedman method, the values of E and A can be determined by the slope and intercept. For each α, a set of E and A value was obtained. Using the slope and the intercept of the linear regression line at each conversion rate in Figure 3, E and A were obtained as shown in the Table 3.

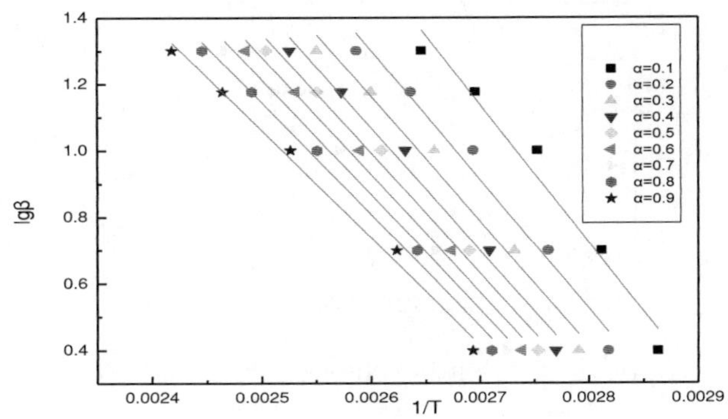

Figure 3. Relationship between $\lg\beta$ and $(1/T)$

Table 3. The values of E and A with the method of the F-W-O.

α	Intercept	Slope	R	E(kJ/mol)
0.1	12.35	4155.63	0.980	75.65
0.2	11.40	3884.86	0.986	70.72
0.3	10.92	3753.21	0.989	68.33
0.4	10.63	3677.23	0.991	66.94
0.5	10.37	3607.14	0.993	65.67
0.6	10.13	3539.35	0.994	64.43
0.7	9.86	3459.17	0.995	62.97
0.8	9.56	3367.33	0.995	61.30
0.9	9.14	3232.98	0.995	58.85

The E obtained by the F-W-O method were averaged to obtain that E_a=66.10kJ/mol.
The Table 4. shows the apparent activation energy in different methods.

Table 4. The values of E with the different methods.

Method	Kissinger	Friedman	F-W-O
E(kJ/mol)	65.23	52.20	66.10

3.5. Determination of curing reaction order.

The reaction order of the curing system can be obtained by the Crane equation.

$$\frac{d(\ln\beta)}{d\left(\frac{1}{T_p}\right)} = -\frac{E}{nR} \tag{4}$$

In equation (4), the physical meanings of β, E, T_p and R are the same as those of equation (1). Reaction order is n. Integrate the equation (4).

$$\ln\beta = -\frac{E}{nR} \times \frac{1}{T_p} + C \tag{5}$$

In equation (5), $\ln\beta$ is linearly related to $1/T_P$. The order of the reaction process is obtained from the slope of the line in Figure 4. According to the equation of Crane, the reaction order of the curing system is obtained by the three methods. The reaction order is shown below.

The calculated reaction order is not an integer, indicating that the curing reaction is a complex reaction [10].

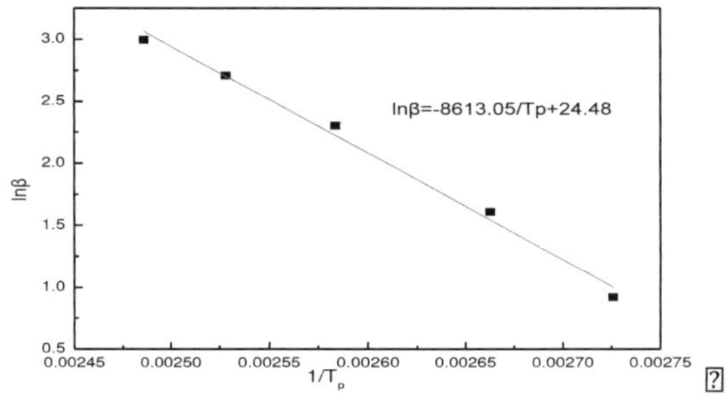

Figure 4. Relationship between $\ln\beta$ and $1/T_P$

Table 5. The values of n with the different methods.

Method	Kissinger	Friedman	F-W-O
n	0.911	0.729	0.923

4. Conclusion

The results show that with the increase of the heating rate, the exothermic peak gradually becomes steep, and the curing time is shortened. The initial temperature, peak temperature and termination temperature of the system are shifted to high temperature.

The curing kinetics of the system was studied by three kinds of analytical methods, and the kinetic apparent activation energy E and the reaction order n were obtained. The results show that the apparent activation energy E of the three methods is different, but it indicates that the system is a complex reaction.

References

[1] Xing S L, Yang J S, et, al. Preparation and characterization of a novel microcapsule-type latent curing agent for epoxy resin[J]. Mater Design, 2015, 18(7): 661-670.

[2] Li S P, Cui C, et al. The effect of hyperbranched polyester and zirconium slag nanoparticles on the impact resistance of epoxy resin thermosets[J]. Compos. Part B: Eng., 2015, 79: 342-350.

[3] Yong L, Wei W, et al. The effects of polyamic acid on curing behavior, thermal stability, and mechanical properties of epoxy/DDS system[J]. Journal of Applied Polymer Science, 2013, 127(4): 3213-3220.

[4] Zhang S T, Wang Z, et al. Study on properties of medium-low temperature cured TGDDM epoxy system with high-performance[J]. Engineering Plastics Application, 2015, 43(3): 40-44.

[5] Zhang J, Guo Q, Fox B. Thermal and mechanical properties of a dendritic hydroxyl-functional hyperbranched polymer and tetra functional epoxy resin blends[J]. Journal of Polymer Science Part: Polymer Physics, 2010, 48(4): 417-424.

[6] Yu J, Zhang B M, et al. Study of cure kinetic model of typical bismelei-mide[J]. Acta Mater Composite Sinica, 2003, 21(1): 78.

[7] Kissinger E D. Reaction kinetics in differential thermal analysis[J]. Anal. Chem., 1957, 29(11): 1702-1706.

[8] Du X J, Li X D, et al. Thermogravimetric analysis and kinetic study of 1-amino-1,2,3-triazolium nitrate[J]. Thermochimica Acta, 2013, 570: 59-63.

[9] Du X J, Li X D, et al. Thermal kinetic study of 1-amino-1,2,3-triazolium nitrate[J]. J Therm Anal Calorim, 2014, 115: 1195-1205.

[10] Xing S L, Wang Z, et al. Study on the curing kinetics of the novel latent epoxy resin systems[J]. Journal of National University of Defense Technology, 2006, 28(2): 31-34.

Investigation on thermal oxidative aging of nitrile rubber (NBR) O-rings under compression stress

X R Liu[1], W F Zhang[1], W T Lou[1], Y X Huang[1] and W Dai[1]

[1] School of Reliability and Systems engineering, Beihang University, Beijing 100191, China

Abstract. The degradation behaviors of nitrile rubber O-rings exposure to air under compression were investigated at three elevated temperatures. The physical and mechanical properties of the aging samples before and after exposure at selected time were studied by measuring weight loss, tensile strength and elongation at break. The Attenuated total reflection Fourier transform infrared (ATR-FTIR) spectroscopy and fracture morphology were used to reveal the microstructural changes of the aging samples. The results indicate that the weight decreased with exposure time and temperature. Based on the results of the crosslinking density, the crosslinking predominates during the most of aging process. The significant changes in tensile strength and elongation at break also indicate the severe degradation in air. The fracture morphology results show that the fracture surface after 64 days of exposure to air turns rough and present defects. The ATR-FTIR results demonstrate that the hydroxyl groups were formed for the samples aged in air.

1. Introduction

NBR is widely used as seals for industrial and domestic applications due to its excellent performance including high resistance to oil, low abrasion rate and compressive stress resistance [1-3]. However, under actual service conditions, the O-rings are often subjected to harsh environment, such as mechanical stress (often the compressive stress), high temperature (up to 150℃), gas atmosphere and liquid environment, leading to the remarkably degradation for the seals [4-7]. Previous reports have focused on the degradation behaviors and mechanisms of nitrile butadiene rubbers in hot air under free state [8-12]. However, in the case of the actual service conditions, the seals are often designed to withstand the compressive stress to achieve the great sealing effect. The compressive stress can also result in the physical and chemical changes of the molecular structure and fillers, and has a direct influence on the degradation behaviors of seals [1]. Until now, just few reports have been available on the degradation of NBR O–rings under the compression state [13, 14]. For example, Zeng et al. [13] studied the corrosion mechanism of hydrogenated nitrile butadiene rubber O-ring under simulated wellbore conditions using self-designed compression simulator and autoclave. Kömmling et al. [14] investigated the effects of heterogeneous aging in compressed NBR and ethylene-propylene-diene rubber (EPDM) O-ring seals. Thus, the degradation behaviors of NBR O–rings exposure to air under compression still need to be further studied.

In this paper, we studied the long-term degradation behaviors of NBR O-rings exposure to air under compression at three aging temperatures by the accelerated testing. Then, the changes in chemical structures of the aged samples were investigated by ATR-FTIR. The physical and mechanical properties of the aged samples were also evaluated by measuring weight loss, mechanical properties and tensile fracture morphology. Then, on the basis of these results, we made an attempt to gain a better understanding of the degradation behaviors of NBR O-rings under the compression state.

2. Materials and methods

2.1. Material and Aging Methods

Nitrile rubber O-rings were supplied by Changsha 5712 Aircraft Industry Corporation, Ltd, China. The main components include NBR, zinc oxide, stearic acid, sulfur, carbon black and additives. An O-ring specimen at an inner diameter of 17.00 mm and a cross-sectional diameter of 2.65 mm. The accelerated aging tests were performed in the air-circulating ovens. In addition, the samples were oven-aged up to 64 days at 70℃, 90℃ and 110℃.

2.2. Characterization Methods

Weight loss of the aged sample was measured by getting the average of the four parallel samples with electronic balance which has a resolution of 0.001 g. Mechanical properties of the aging samples were estimated by tensile tests. The tensile samples were tested according to GB/T 5720:1993 (China) using CMT5504 electronic testing machine at a strain rate of 50 mm/min. The tensile fracture morphology changes of the aging samples were analyzed by using scanning electron microscopy (SEM, JSM-6480). Attenuated total reflection-Fourier transform infrared (ATR-FTIR, Thermo Nicolet 6700) spectroscopy was applied to study the changes of chemical structure of the seals surfaces. The range of spectra is 400-4000 $cm^{-1}$with the resolution of 4 cm^{-1}.

3. Results and discussion

3.1. Weight loss

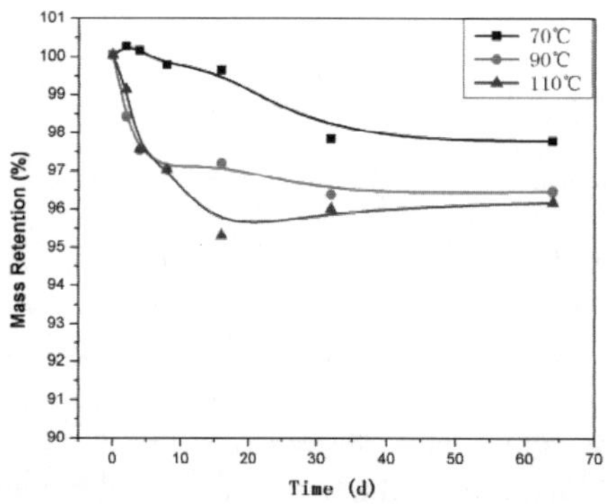

Fig 1. Weight loss of the compressed samples with the exposure time at different temperatures.

The changes in weight loss of the compressed samples at different temperatures are shown in fig 1. The weight loss increased with exposure time at 70℃. And the changes in weight show a similar trend at 90℃ and 110℃. The weight loss increased remarkably during the first 4 days and then kept stable after 16 days of exposure. The higher the temperature, the more the weight loss. When the O-rings were exposed to air at high temperature, the volatile additives and residual monomers gradually migrated from the inner to surface by the action of elevated temperature, and then volatilized in atmosphere, resulting in the loss of the weight [10].

3.2. Mechanical properties

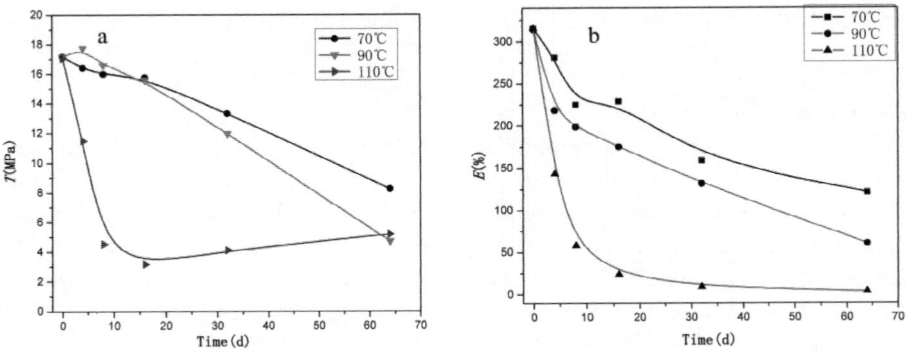

Fig 2. Changes in mechanical properties of the aging samples with the exposure time at different temperatures: a-tensile strength; b-elongation at break.

When the O-rings were subjected to the compressive stress, see fig 2, in the early stage, the stress caused the rearrangement of molecular chains accompanied by the decrease in free volume between the chains and the worse compatibility between the fillers and polymer matrix, which resulted in the slightly decrease of the tensile strength. However, in the early stage of aging at 90℃, the appropriate crosslinking causes the increase in the tensile strength. For the compressed samples exposure to air, the consumptions of oxygen can meet the demand of the crosslinking reactions, resulting in the obvious increase of the crosslinking density and the formation of the denser network structure, which lead to the decrease in the tensile strength. But the limitation of oxygen diffusion, the stress and the elevated temperature cause the decrease in the crosslinking density and seriously destroy the matrix structure, leading to the decrease in the tensile strength. Meanwhile, the stress-induced and oxidation-induced physical and chemical degradations both result in the decrease in the elongation at break. These results demonstrate that the compressive stress and temperature have a great effect on the mechanical properties of the NBR O-rings.

3.3. Fracture morphology

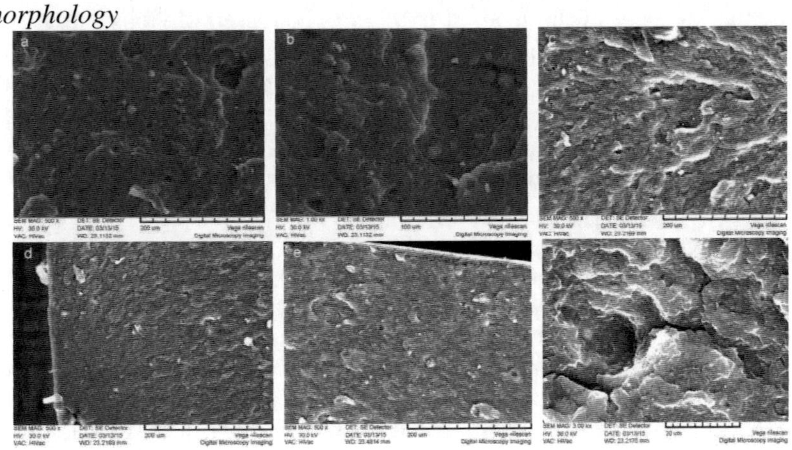

Fig 3. SEM of the tensile fracture surface of the aging samples at 110℃ for different time.

The fracture morphology of the compressed samples before and after exposure to air at 110℃ for different time is shown in fig 3. It is clearly seen that the obvious changes occurred in the fracture surface over time compared to the relatively smooth and uniform distribution of fillers fracture surface for the unaged samples. For the compressed samples after 64 days of exposure to air, the fracture surface (shown in fig 3 (c)) became rough and presented many secondary cracks, implying the characteristics of brittle fracture. Fig 3 (d) and fig 3 (e) show the formation of the hardened brittle outer layer. It is obvious that the hardened brittle outer layer contact with flanges is less pronounced than that exposure to air, implying the limitation of oxygen diffusion caused by flanges and the more serious degradation for the samples exposure to air. The fracture surface at a magnification of 3000X shows network crack distributes desultorily in the polymer matrix, indicating the great changes for the network structure occurred.

3.4. ATR-FTIR analysis

To elevate whether the chemical degradation of the compressed samples exposure to air happened, the ATR-FTIR spectrum of the aged samples were investigated. Fig 4 shows the ATR-FTIR spectra of the compressed samples before and after exposure to air at 70℃and 110℃, respectively. The peak at 3394 cm^{-1} is assigned to the stretching vibration of the O-H band, implying the formation of the hydroxyl group. The peaks at 2918 cm^{-1} and 2843 cm^{-1} are attributed to the $C-H$ stretching vibration of the CH_2 group. The peak at 2232 cm^{-1} is assigned to the $-CN$ stretching vibration. The peaks at 1733 cm^{-1} and 1533 cm^{-1} correspond to the $C=O$ stretching vibration, indicating the presence of additives such as stearic acid and curing package. The peak at 1433 cm^{-1} is from the $C-H$ bending vibration of the CH_2 group. The peak at 962 cm^{-1} is attributed to the $C-H$ deformation vibration of the trans-1, 4-structure of the butadiene part.

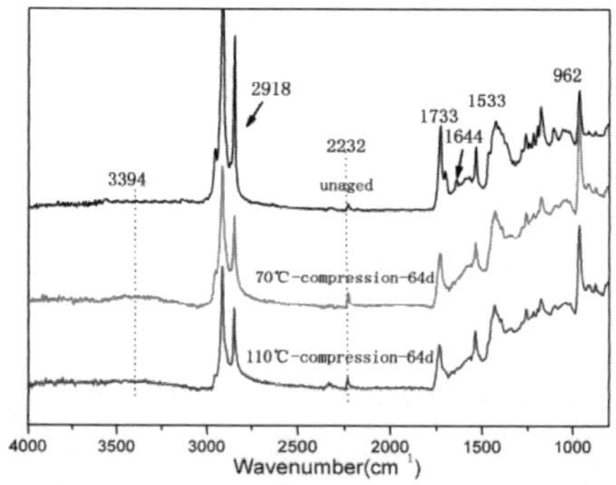

Fig 4. ATR-FTIR spectra of the aging samples with the exposure time at 70℃and 110℃.

By comparing the spectrum of the compressed samples after exposure to air at 70℃ and 110℃, it can be clearly seen that the intensity of the peak at 3394 cm^{-1} increased obviously for the aged samples with aging time, which indicated that hydroxyl groups were generated due to the oxidation of the samples during the degradation process. For the samples after exposure to air, the peak at 1733 cm^{-1} first increased and then decreased with increase in aging time, implying the formation and consumption of carboxyl groups. It can be concluded that the carboxyl groups and hydroxyl groups were formed for the compressed samples exposure to air during the aging process.

4. Conclusions

The degradation of the NBR O-rings under the compression state exposure to air at elevated temperature were studied. It can be concluded that, during the aging process, the weight loss increased sharply first and then remained constant with exposure time and temperature, indicating the migration of fillers. The changes in tensile strength and elongation at break implying the serious degradation for the compressed samples exposure to air. The fracture morphology shows that the defects (voids and agglomerates) and the hardened brittle outer layer were generated in the fracture surface. The ATR-FTIR results indicate that hydroxyl groups and carboxyl groups were formed for the samples aged in air.

References

[1] Akhlaghi S, Gedde U W, Hedenqvist M S, Braña M T C and Bellander M 2015 *Renew. Sust. Energ. Rev.* **43** 1238–48
[2] Guo F, Jia X, Lv M, Wang L, Salant R F and Wang Y 2014 *Tribol. Int.* **78** 187–94
[3] Pazur R J, Cormier J G and Korhan-Taymaz K 2014 *Rubber. Chem. Technol.* **87** 239–49
[4] Liu J, Li X, Xu L and Zhang P 2016 *Polym. Test* **54** 59–66
[5] Delor-Jestin F, Barrois-Oudin N, Cardinet C, Lacoste J and Lemaire J 2000 *Polym. Degrad. Stabil.* **70** 1–4
[6] Zhao J, Yang R, Iervolino R and Barbera S 2013 *Rubber. Chem. Technol.* **86** 591–603

[7] Buckley G S, Roland C M 2014 *J. Appl. Polym. Sci.* **131** 2928–35
[8] Alves S M, Mello V S and Medeiros J S 2013 *Tribol. Int.* **65** 74–80
[9] Coronado M, Montero G, Valdez B, Stoytcheva M, Eliezer A, García C, Campbell H and Pérez A 2014 *Energ.* **68** 364–69
[10] Akhlaghi S, Hedenqvist M S, Conde Braña M T, Bellander M and Gedde U W 2015 *Polym. Degrad. Stabil.* **111** 211–22
[11] Pan L, Tan J, Han X, Li P and Zhang W 2016 *J. Appl. Polym. Sci.* **133**
[12] Alcock B, Jørgensen J K 2015 *Polym. Test* **46** 50–8
[13] Zeng D, He Q, Li T, Hu J, Shi T, Zhang Z, Yu Z and Liu R 2016 *Corros. Sci.* **107** 145–54
[14] Kömmling A, Jaunich M and Wolff D 2016 *Polym. Degrad. Stabil.* **126** 39–46

Effective antiplane shear moduli of the composites with doubly periodic fibers

Yao-Ling Xu[1] and Yan-Zhi Shen[2]

[1] Key Laboratory of Mechanical Reliability for Heavy Equipments and Large Structures of Hebei Province, Yanshan University, Qinhuangdao 066004, P.R.China
[2] College of Mechanics Engineering, Yanshan University, Qinhuangdao 066004, P.R.China

xylysu@163.com

Abstract. A novel method for determining the elastic field in the composite with doubly periodic fibers under far-field antiplane strains is presented. The composite is replaced by a homogeneous medium with doubly periodic stresses, which are not related to the strains, in the regions corresponding to the fibers of the composite, the equivalence condition between the composite and homogeneous medium is established. The homogeneous medium with the doubly periodic stresses is solved, and the elastic fields are obtained in the doubly periodic fibers and the matrix. The obtained elastic field is used to evaluate the effective antiplane shear moduli of the composites, good agreements with the existing results are observed.

1. Introduction

Due to weight reducing, strengthening, better fatigue strength and improved corrosion resistance compared to conventional materials, fiber reinforced composites are widely-used in engineerings. The composites are heterogeneous medium, which can be seen as homogeneous mediums in structural analysis, it is great important to design and evaluate the effective material properties of such composites. The composite model with periodically distributed reinforcements is certified to be an excellent model for the simulation of effective properties of the composites [1]. Different analytical methods have been developed to study composites with periodic microstructures, such as the Fourier series expansion method [2], the asymptotic homogenization method [3,4], eigenfunction expansion-variational method [5]. Also, the numerical methods such as the boundary element method [6] and the finite element method [7,8] are effective methods to study the composites with periodic microstructures. This paper presents a novel method for determining the elastic field of the composites containing doubly periodic fibers under far-field antiplane strains. The effective antiplane shear moduli of the composites are evaluated and compared with the existing results.

2. Statement of the problem

As shown in figure 1, the parallelogram P_{00} denotes the fundamental cell of the periodic composites, there is a circular cross-section fiber with radius R in the P_{00}, the fiber and the matrix are assumed to be perfectly bonded, the boundary of the P_{00} is denoted by $\Gamma = \Gamma_1 + \Gamma_2 + \Gamma_3 + \Gamma_4$, Ω_f^0 and Ω_m^0 denote the regions occupied by the fiber bounded by the contour L_0 and matrix, respectively. The fundamental cell P_{00} is periodically arranged in the complex plane $z = x_1 + \mathrm{i}x_2$, the cross-section of the

IOP Conf. Series: Materials Science and Engineering **265** (2017) 012004 doi:10.1088/1757-899X/265/1/012004

composite is also shown in figure 1. Let Ω_f and Ω_m denote the union of Ω_f^0 and Ω_m^0 and its periodic congruent regions respectively, L denotes the union of L_0 and its periodic congruent contours. $2\omega_1$ and $2\omega_2$ are two fundamental periods, the vertices of the P_{00} are $\omega_1+\omega_2$, $-\omega_1+\omega_2$, $-\omega_1-\omega_2$ and $\omega_1-\omega_2$, respectively. The shear moduli of the matrix and the fibers are G_m and G_f, respectively. The composite is subjected to the far-field antiplane strains γ_{13}^{∞} and γ_{23}^{∞}.

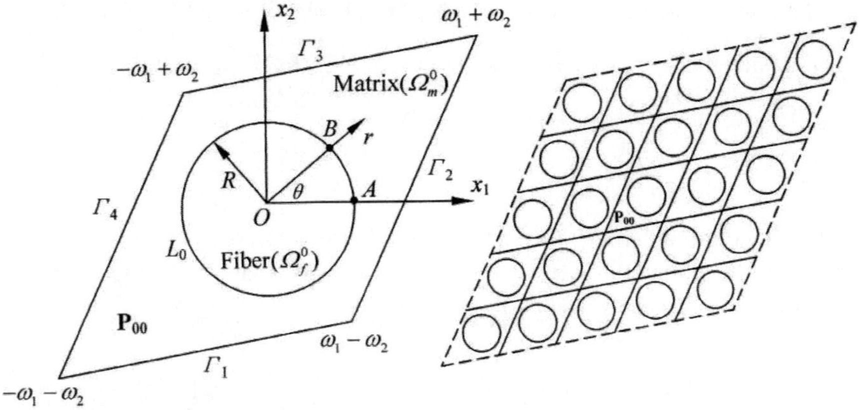

Figure 1. Fundamental cell P_{00} and cross section of the composites

3. Homogenization of heterogeneous medium

For the antiplane problem, there are only the non-trivial antiplane displacement w, strain components γ_{13} and γ_{23}, stress components τ_{13} and τ_{23}, and all field quantities being only the functions of coordinates x_1 and x_2. Let $\tau=\tau_{13}-\mathrm{i}\tau_{23}$ be complex stress, and $\gamma=\gamma_{13}-\mathrm{i}\gamma_{23}$ be complex strain.

The original problem: an infinite homogeneous medium with the shear modulus G_m subjected to a far-field antiplane strain $\gamma^{\infty}=\gamma_{13}^{\infty}-\mathrm{i}\gamma_{23}^{\infty}$. Apparently, in the entire plane, the strain field γ^0 and the stress field τ^0 are uniform, and given by $\gamma^0=\gamma^{\infty}$ and $\tau^0=G_m\gamma^0=G_m\gamma^{\infty}$. For an actual heterogeneous medium under consideration, the presence of the doubly periodic fibers disturbs the uniform strain and stress field. Denote the disturbance strains and stresses by γ_f' and τ_f' in Ω_f and γ_m' and τ_m' in Ω_m respectively, the total stresses and strains in Ω_f and Ω_m are

$$\tau^0+\tau_f'=G_m\gamma^{\infty}+\tau_f' \qquad \text{in } \Omega_f \qquad (1)$$

$$\gamma^0+\gamma_f'=(\tau^0+\tau_f')/G_f=(G_m\gamma^{\infty}+\tau_f')/G_f \qquad \text{in } \Omega_f \qquad (2)$$

$$\tau^0+\tau_m'=G_m\gamma^{\infty}+\tau_m' \qquad \text{in } \Omega_m \qquad (3)$$

$$\gamma^0+\gamma_m'=(\tau^0+\tau_m')/G_m=(G_m\gamma^{\infty}+\tau_m')/G_m \qquad \text{in } \Omega_f \qquad (4)$$

The equivalent problem: an infinite homogeneous medium with the shear modulus G_m subjected to a far-field antiplane strain $\gamma^{\infty}=\gamma_{13}^{\infty}-\mathrm{i}\gamma_{23}^{\infty}$. In the entire plane, the strain field γ^0 and the stress field τ^0 are $\gamma^0=\gamma^{\infty}$ and $\tau^0=G_m\gamma^0=G_m\gamma^{\infty}$ respectively. Instead of dealing with the presence of the doubly periodic fibers in the original problem, doubly periodic stresses $\tau^*=\tau_{13}^*-\mathrm{i}\tau_{23}^*$, which are not related to the strains, are introduced in Ω_f. The introduction of the stresses disturbs the uniform strain and stress fields, with the disturbance strains and stresses being denoted by $\gamma_{in}'^*$ and $\tau_{in}'^*$ in Ω_f and $\gamma_{out}'^*$ and $\tau_{out}'^*$ in Ω_m, the total stresses and strains are

$$\tau^0 + \tau^* + \tau_{in}^{'*} = G_m\gamma^\infty + \tau^* + \tau_{in}^{'*} \qquad \text{in } \Omega_f \tag{5}$$

$$\gamma^0 + \gamma_{in}^{'*} = (G_m\gamma^\infty + \tau^* + \tau_{in}^{'*} - \tau^*)/G_m = (G_m\gamma^\infty + \tau_{in}^{'*})/G_m \qquad \text{in } \Omega_f \tag{6}$$

$$\tau^0 + \tau_{out}^{'*} = G_m\gamma^\infty + \tau_{out}^{'*} \qquad \text{in } \Omega_m \tag{7}$$

$$\gamma^0 + \gamma_{out}^{'*} = (\tau^0 + \tau_{out}^{'*})/G_m = (G_m\gamma^\infty + \tau_{out}^{'*})/G_m \qquad \text{in } \Omega_m \tag{8}$$

The equivalence between the two problems requires the same stresses and strains, let

$$\tau^* + \tau_{in}^{'*} = \tau_f^{'} \tag{9}$$

$$\tau_{out}^{'*} = \tau_m^{'} \tag{10}$$

and from equations (1), (5) and (3), (7), the stress equivalence is arrived at. A comparison of equations (4) and (8) shows that the strain equivalence in Ω_m is also arrived at. From equations (2) and (6), the strain equivalence in Ω_f results in the following equation

$$(\frac{G_m}{G_f} - 1)\gamma^\infty + (\frac{1}{G_f} - \frac{1}{G_m})\tau_{in}^{'*} + \frac{1}{G_f}\tau^* = 0 \tag{11}$$

4. Elastic field induced by the stress τ^*

The analytical function $F(z)$ can be used to formulate the antiplane problem:

$$w = F(z) + \overline{F(z)} \tag{12}$$

$$\tau = G_m\gamma = 2G_m\, dF(z)/dz \tag{13}$$

$$T = \int_A^B \tau_{nz} ds = \int_A^B (\tau_{13}dx_2 - \tau_{23}dx_1) = G_m \mathrm{i}[\overline{F(z)} - F(z)]_A^B \tag{14}$$

where T is the resultant shear stress on an arc AB, $[\,\cdot\,]_A^B$ signifies the change in the bracketed function in moving from point A to point B along the arc AB, the overbar denotes the conjugate.
To describe the stress τ^*, a function $F^*(z)$ is introduced and expanded into Taylor series in Ω_f^0

$$F^*(z) = \sum_{k=1}^{\infty} A_k z^k \qquad z \in \Omega_f^0 \tag{15}$$

From equations (13) and (15), the stress τ^* can be expressed as

$$\tau^* = \tau_{13}^*(z) - \mathrm{i}\tau_{23}^*(z) = 2G_m \frac{dF^*(z)}{dz} = 2G_m \sum_{k=1}^{\infty} kA_k z^{k-1} \qquad z \in \Omega_f^0 \tag{16}$$

The stress on boundary L_0 in polar coordinates (r,θ) can be expressed as

$$\tau_{r3}^*(t) - \mathrm{i}\tau_{\theta 3}^*(t) = e^{\mathrm{i}\theta}[\tau_{13}^*(t) - \mathrm{i}\tau_{23}^*(t)] = \frac{2G_m}{R}\sum_{k=1}^{\infty} kA_k t^k \qquad t \in L_0 \tag{17}$$

The interfacial stress $\tau_{r3}^*(t)$ on L_0 is

$$\tau_{r3}^*(t) = \frac{[\tau_{r3}^*(t) - \mathrm{i}\tau_{\theta 3}^*(t)] + \overline{[\tau_{r3}^*(t) - \mathrm{i}\tau_{\theta 3}^*(t)]}}{2} = \frac{G_m}{R}(\sum_{k=1}^{\infty} kA_k t^k + \sum_{k=1}^{\infty} k\bar{A}_k R^{2k}\frac{1}{t^k}) \qquad t \in L_0 \tag{18}$$

Selecting arc AB on L_0 as shown in figure 1, A is a fixed point on x_1 axis and $\boldsymbol{B}$ is a moving point. The resultant shear stress on arc AB can be written as

$$T^*(t)=\int_A^B \tau_{r3}^*(t)ds \qquad t\in L_0 \qquad (19)$$

The substitution of equation (18) into equation (19) yields

$$T^*(t)=\frac{G_m}{\mathrm{i}}\sum_{k=1}^{\infty}[(A_k t^k-\bar{A}_k R^{2k}\frac{1}{t^k})-R^k(A_k-\bar{A}_k)] \qquad t\in L_0 \qquad (20)$$

Let $F(z)$ denotes the complex potential function of the elastic field induced by the stress τ^*. Owing to the double periodicity of the stress τ^*, the stresses and strains induced by τ^* are also doubly periodic, the displacement field induced by τ^* and the function $F(z)$ are doubly quasi-periodic.
The continuity conditions of the displacement on L can be written as

$$w^+(t)-w^-(t)=0 \qquad t\in L \qquad (21)$$

from equation (12), equation (21) can be rewritten as

$$[F^+(t)+\overline{F^+(t)}]-[F^-(t)+\overline{F^-(t)}]=0 \qquad t\in L \qquad (22)$$

The jump conditions of the resultant on L can be written as

$$T^+(t)-T^-(t)=-T^*(t) \qquad t\in L \qquad (23)$$

from equation (14), equation (23) can be rewritten as

$$[F^+(t)-\overline{F^+(t)}]-[F^-(t)-\overline{F^-(t)}]=\frac{T^*(t)}{G_m\mathrm{i}} \qquad t\in L \qquad (24)$$

where the superscripts "+" and "-" signify the boundary values of the quantities as approached from the interior and the exterior regions of the contour L, respectively.
From equations (22) and (24), it is seen that

$$F^+(t)-F^-(t)=\frac{T^*(t)}{2G_m\mathrm{i}} \qquad t\in L \qquad (25)$$

The solution of equation (25) (a doubly quasi-periodic Riemann boundary problem) in the P_{00} is

$$F(z)=C_0 z+\frac{1}{2\pi i}\int_{L_0}\frac{T^*(t)}{2G_m\mathrm{i}}\zeta(t-z)dt \qquad z\in\Omega_f^0+\Omega_m^0 \qquad (26)$$

where $\zeta(\cdot)$ is Weierstrass Zeta function, C_0 is the constant to be determined.
Although $F(z)$ is solved in the fundamental cell P_{00}, it only differs by an addend in the neighboring unit cell, which is not related to the stress and strain. Substituting equation (20) into equation (26), $F(z)$ can be determined as following

$$F(z)=C_0 z-\frac{1}{2}\sum_{k=1}^{\infty}\left\{A_k z^k-\frac{(-1)^k}{(k-1)!}\bar{A}_k R^{2k}[\zeta(z)-\frac{1}{z}]^{(k-1)}-R^k(A_k-\bar{A}_k)\right\} \qquad z\in\Omega_f^0 \qquad (27)$$

$$F(z)=C_0 z+\frac{1}{2}\sum_{k=1}^{\infty}\frac{(-1)^k}{(k-1)!}\bar{A}_k R^{2k}\zeta^{(k-1)}(z) \qquad z\in\Omega_m^0 \qquad (28)$$

then the strains and stresses induced by the stress τ^* are derived from equation (13)

$$\gamma_{in}^{'*} = 2C_0 - \sum_{k=1}^{\infty}\left\{kA_k z^{k-1} - \frac{(-1)^k}{(k-1)!}\bar{A}_k R^{2k}[\zeta(z) - \frac{1}{z}]^{(k)}\right\} \quad z \in \Omega_f^0 \tag{29}$$

$$\gamma_{out}^{'*} = 2C_0 + \sum_{k=1}^{\infty}\left\{\frac{(-1)^k}{(k-1)!}\bar{A}_k R^{2k}\zeta^{(k)}(z)\right\} \quad z \in \Omega_m^0 \tag{30}$$

$$\tau_{in}^{'*} = 2G_m C_0 - \sum_{k=1}^{\infty} G_m\left\{kA_k z^{k-1} - \frac{(-1)^k}{(k-1)!}\bar{A}_k R^{2k}[\zeta(z) - \frac{1}{z}]^{(k)}\right\} \quad z \in \Omega_f^0 \tag{31}$$

$$\tau_{out}^{'*} = 2G_m C_0 + \sum_{k=1}^{\infty} G_m\left\{\frac{(-1)^k}{(k-1)!}\bar{A}_k R^{2k}\zeta^{(k)}(z)\right\} \quad z \in \Omega_m^0 \tag{32}$$

For far-field antiplane strain γ^{∞}, owing to the periodicity, the difference of the disturbed displacement between the two endpoints of each boundary Γ_k is zero. From equation (12), it follows that

$$[\overline{F(z)} + F(z)]_{\Gamma_k} = 0 \tag{33}$$

The substitution of equation (28) into equation (33) yields

$$C_0 = \frac{\pi R^2}{2S}(A_1 + \delta_2 \bar{A}_1) \tag{34}$$

where $S = 2\mathrm{i}(\omega_1\bar{\omega}_2 - \omega_2\bar{\omega}_1)$ is the area of the P_{00}, $\delta_2 = 2(\bar{\omega}_1\eta_2 - \bar{\omega}_2\eta_1)/(\pi \mathrm{i})$, $\eta_1 = \zeta(\omega_1)$, $\eta_2 = \zeta(\omega_2)$.

5. Stresses and strains in the fibers and the matrix

Noting equations (9) and (10), from equations (1-4), the total stresses and strains are

$$\tau_f = G_m\gamma^{\infty} + \tau^{*} + \tau_{in}^{'*} = \mathrm{G_m}\gamma^{\infty} + 2G_m\sum_{k=1}^{\infty} kA_k z^{k-1} + G_m\left\{\frac{\pi R^2}{S}(A_1 + \delta_2\bar{A}_1) - \sum_{k=1}^{\infty}\left[kA_k z^{k-1} - \frac{(-1)^k}{(k-1)!}\bar{A}_k R^{2k}[\zeta(z) - \frac{1}{z}]^{(k)}\right]\right\} \quad z \in \Omega_f^0 \tag{35}$$

$$\gamma_f = \tau_f / G_f \quad z \in \Omega_f^0 \tag{36}$$

$$\tau_m = G_m\gamma^{\infty} + \tau_{out}^{'*} = G_m\gamma^{\infty} + G_m\left\{\frac{\pi R^2}{S}(A_1 + \delta_2\bar{A}_1) + \sum_{k=1}^{\infty}\left[\frac{(-1)^k}{(k-1)!}\bar{A}_k R^{2k}\zeta^{(k)}(z)\right]\right\} \quad z \in \Omega_m^0 \tag{37}$$

$$\gamma_m = \tau_m / G_m \quad z \in \Omega_m^0 \tag{38}$$

Substituting equations (16) and (31) in equation (11), the equivalent condition can be written as

$$(\frac{G_m}{G_f} - 1)\gamma^{\infty} + (\frac{G_m}{G_f} - 1)\left\{\frac{\pi R^2}{S}(A_1 + \delta_2\bar{A}_1) - \sum_{k=1}^{\infty}\left[kA_k z^{k-1} - \frac{(-1)^k}{(k-1)!}\bar{A}_k R^{2k}[\zeta(z) - \frac{1}{z}]^{(k)}\right]\right\} + 2\frac{G_m}{G_f}\sum_{k=1}^{\infty} kA_k z^{k-1} = 0 \tag{39}$$

From equation (39), the constants A_k (k=1,2,3,...) can be uniquely determined, then the stresses and strains in the doubly periodic fibers and the matrix can be obtained from equations (35-38).

6. Effective antiplane moduli of composites

According to the average strain theorem of elasticity, for a two-phase composite under far-field antiplane strain $\gamma^{\infty} = \gamma_{13}^{\infty} - i\gamma_{23}^{\infty}$, the effective antiplane moduli are determined by the following equation

$$\begin{bmatrix} G_m - C_{44}^{eff} & -C_{45}^{eff} \\ -C_{45}^{eff} & G_m - C_{55}^{eff} \end{bmatrix} \begin{Bmatrix} \gamma_{23}^{\infty} \\ \gamma_{13}^{\infty} \end{Bmatrix} = \lambda \begin{bmatrix} G_m - G_f & 0 \\ 0 & G_m - G_f \end{bmatrix} \begin{Bmatrix} \bar{\gamma}_{23,f} \\ \bar{\gamma}_{13,f} \end{Bmatrix} \tag{40}$$

where λ is fiber volume fraction, $\bar{\gamma}_{13,f}$ and $\bar{\gamma}_{23,f}$ are the averaged antiplane strains.

Fiber square arrangement bring out transverse isotropy, i.e., $C_{44}^{eff} = C_{55}^{eff} = G^{eff}$, $C_{45}^{eff} = 0$, the effective antiplane shear moduli of the square fiber arrangement was studied by Fourier series expansion method [2], a comparison with literature [2] is shown in table 1, good agreements are observed.

Table 1. A comparison with the Fourier series expansion method

λ	G_f/G_m	G^{eff}/G_m	
		literature [2]	present
0.4	6	1.805	1.80451
	20	2.147	2.14586
	120	2.314	2.31343
0.55	6	2.326	2.32562
	20	3.184	3.07712
	120	3.555	3.50663
0.7	6	3.176	3.17311
	20	5.222	5.21367
	120	6.945	6.92944
0.75	6	3.620	3.61966
	20	7.006	7.00409
	120	11.170	11.1640

7. Conclusions

Using the equivalent inclusion technique integrated with the result of doubly periodic Riemann boundary value problem, an analytic method for determining the elastic field of the composites with doubly periodic fibers under far-field antiplane strains is developed. The solution is used to evaluate the effective antiplane moduli of the composites, comparisons with the existing results show the correctness of the present solution. The present method can provide benchmark results for other methods, and may be useful in the computer-aid design of new materials.

References

[1] Trias D, Costa J, Mayugo J A and Hurtado J E 2006 *Comp. Mater. Sci.* **38** 316–24
[2] Chen C H 1970 *J. Appl. Mech. –T. Asme* **37** 198–201
[3] Rodriguez-Ramos R, Sabina F J, Guinovart-Diaz R and Bravo-Castillero J 2001 *Mech. Mater.* **33** 223–35
[4] Andrianov I V, Danishevs'kyy V V and Kalamkarov A L 2008 *Composites Part B* **39** 874–81
[5] Yan P, Jiang C P and Song F 2011 *Mech. Mater.* **43** 586–97
[6] Dong C Y 2006 *Int. J. Solids Struct.* **43** 7919–38
[7] Xia Z H, Zhang Y F and Ellyin F 2003 *Int. J. Solids Struct.* **40** 1907–21
[8] Xu Y L, Du S S, Xiao J H and Zhao Q X 2012 *Comp. Mater. Sci.* **61** 34–41

IOP Conf. Series: Materials Science and Engineering **265** (2017) 012005 doi:10.1088/1757-899X/265/1/012005

Magnetic materials for mobile communication antennas substrate application

Hui Chen[1,*], Difei Liang[1,2], Weijia Li[1] and Chao Pang[1]
[1] National Engineering Research Center of Electromagnetic Radiation Control Materials, University of Electronic Science and Technology of China
[2] State Key Laboratory of Electronic Thin Films and Integrated Devices

E-mail address: 1373925409@qq.com

Abstract. In this work, 3Ba0.7Sr0.3O · 2CoO · 10.8Fe2O3 and Ba2Co2Fe12O22 had been fabricated successfully by conventional ceramic process. Crystallographic structure and electromagnetic properties of two kind of hexagonal ferrite with different sintering temperature were investigated. X-ray Diffraction (XRD), Agilent-N5230A Network Analyzer were used to measure ferrite samples. The mobile phone antenna performance was analysed by HFSS. The results revealed that the main phase of two ferrite samples generated at lower temperature due to additive. The optimized parameters of ferrite are sintering temperature at 1000°C. And to emulate antenna model by HFSS find that Z-type and Y-type ferrite substrate can contribute to antenna frequency shifting, radiation efficiency were affected a little.

1. Introduction

As the miniaturization and high frequency of mobile phones, the size of antenna needs to be reduced. According to the equation of $\lambda = \frac{c}{f_r\sqrt{\mu_{eff} \times \varepsilon_{eff}}}$ and $Z = \sqrt{\mu/\varepsilon} = \eta_0\sqrt{\mu'/\varepsilon'}$, increasing the permeability and permittivity of medium can reduce physical dimension of the antenna. And when permeability (μ) is equal to permittivity (ε), characteristic impedance of the antenna is match to that of free space. Absorption and radiation efficiency (RE) of antenna can become better.

In this paper, in order to reduce the size of antenna, Z-type and Y-type hexagonal ferrite are studied. Using ferrite as the substrate of antenna, ferrite possesses both permeability (μ) and permittivity (ε), it can move the resonant frequency of antenna to lower frequency, so we can design smaller antenna. Z-type and Y-type hexagonal ferrite have high resonant frequency, there are many research of them under 900MHz for mobile antenna application in recent years. We investigate the effects of sintering temperature on electromagnetic properties of Co_2Z and Co_2Y at GHz in this paper.

2. Experimental

$3Ba_{0.7}Sr_{0.3}O \cdot 2CoO \cdot 10.8Fe_2O_3(Co_2Z)$ and $Ba_2Co_2Fe_{12}O_{22}(Co_2Y)$ are synthesized through conventional ceramic techniques, with Fe_2O_3 (99.54wt%) 、 $BaCO_3$ (99wt%) 、 Co_2O_3 （99wt%）, and $SrCO_3$ （99wt%）. The raw materials were mixed, dried and calcined at 1240°C for Co_2Z and 1200°C for Co_2Y, both Z and Y calcined powders were ground by a ball mill with 3wt% Bi_2O_3, after drying, the powder were granulated with polyvinyl alcohol (PVA) and pressed to shape of

toroid. The toroidal samples were sintered at 1000°C, 1100°C, 1200°C for Z and 900°C, 950°C, 1000°C, 1050°C for Y.
Crystallographic structure were analysed by using X-ray Diffraction (XRD). Agilent-N5230A Network Analyzer was used to obtain complex permeability and complex permittivity in the frequency from 0.5GHz to 4GHz. S11 and RE of antenna were simulated by HFSS.

3. Results and discussion

The XRD patterns of $3Ba_{0.7}Sr_{0.3}O \cdot 2CoO \cdot 10.8Fe_2O_3$($Co_2Z$) sintered at 1000°C, 1100°C, 1200°C and $Ba_2Co_2Fe_{12}O_{22}$ (Co_2Y) sintered at 900°C, 950°C, 1000°C, 1050°C are shown in Fig.1, all of the samples have been verified to be a primary Z-type hexagonal phase with a minor non-magnetic $BaFe_2O_4$ phase, when sintering temperature decreases from 1200°C to 1000°C, the more Y-type hexagonal phase appear. Z-type phase dose not form through once process by those oxides (Hongguo Zhang et al.,2001). It produces M-type ($BaFe_{12}O_{19}$) phase and Y-type phase first, then, Z-type phase starts to form as the temperature increases. In the experiment, the additive of 3wt% Bi_2O_3 was used in the second ball mill, it can reduce the sintering temperature. So, the main Z-type phase can form at 1000°C. All of the Y-type samples have been verified to be a primary Y-type hexagonal phase, this confirms that changing the sintering temperature can't give any other phases.

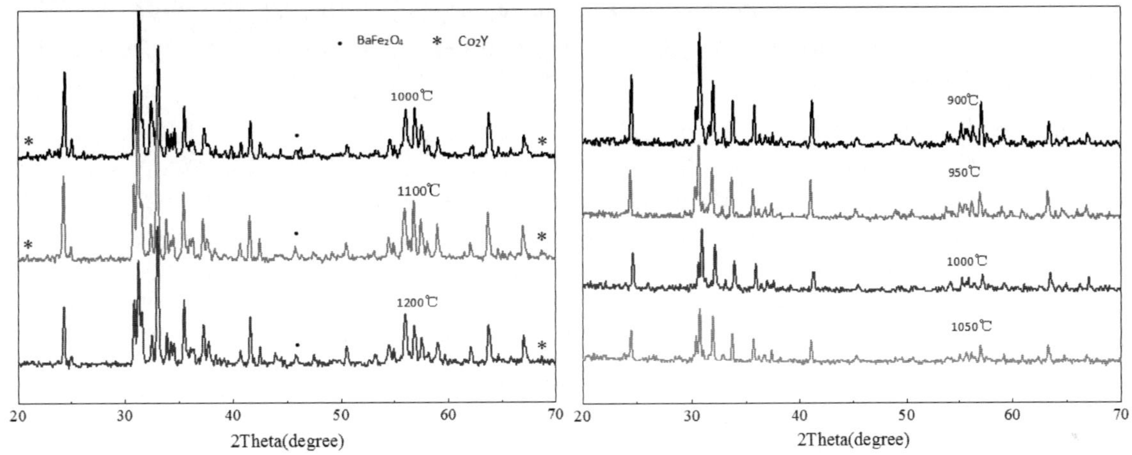

(a) (b)

Figure 1. XRD patterns for (a) $3Ba_{0.7}Sr_{0.3}O \cdot 2CoO \cdot 10.8Fe_2O_3$ ferrite at 1000°C, 1100°C and 1200°C, (b) $Ba_2Co_2Fe_{12}O_{22}$ ferrite sintered at 900°C, 950°C, 1000°C and 1050°C.

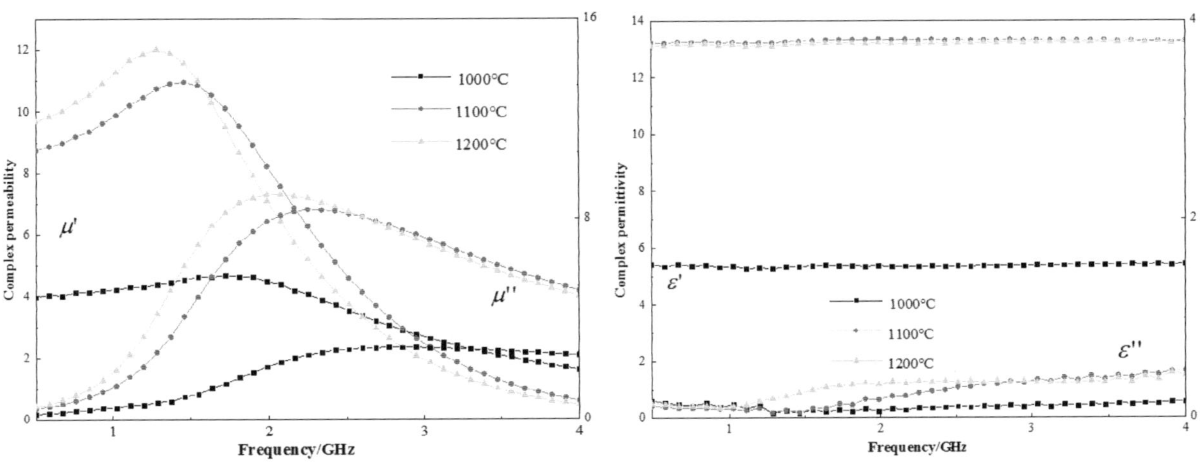

(a) (b)

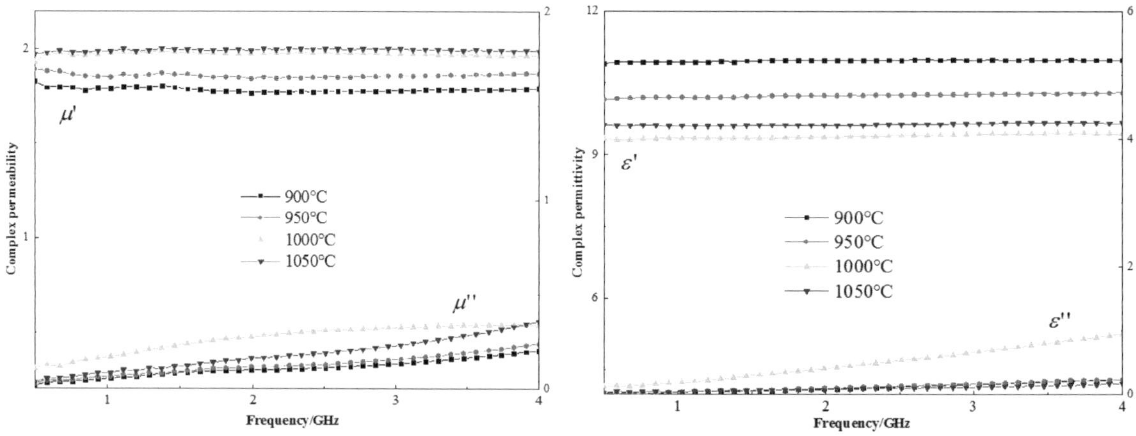

(c) (d)

Figure2. Complex permeability spectra for (a) $3Ba_{0.7}Sr_{0.3}O \cdot 2CoO \cdot 10.8Fe_2O_3$ ferrite at 1000°C, 1100°C and 1200°C; (c) $Ba_2Co_2Fe_{12}O_{22}$ ferrite sintered at 900°C, 950°C, 1000°C and 1050°C. Complex permittivity spectra for (b) $3Ba_{0.7}Sr_{0.3}O \cdot 2CoO \cdot 10.8Fe_2O_3$ ferrite at 1000°C, 1100°C and 1200°C; (d) $Ba_2Co_2Fe_{12}O_{22}$ ferrite sintered at 900°C, 950°C, 1000°C and 1050°C.

Fig.2 indicate the permeability and permittivity changes of $3Ba_{0.7}Sr_{0.3}O \cdot 2CoO \cdot 10.8Fe_2O_3$ and $Ba_2Co_2Fe_{12}O_{22}$ (Co_2Y) ferrite with the sintering temperature. Both the real and imaginary permeability of $3Ba_{0.7}Sr_{0.3}O \cdot 2CoO \cdot 10.8Fe_2O_3$ increase at 1GHz, the peaks of μ'' transfer to a lower frequency with the increasing of sintering temperature and the resonant frequency at GHz. We can see the permeability and permittivity of $Ba_2Co_2Fe_{12}O_{22}$ remain nearly constant at all the frequency of 0.5GHz to 4GHz. It can be predicted the phase of Y-type has already formed at 900°C, it's due to Bi_2O_3 additive. It also can be state that Y-type ferrite has a very high resonant frequency in this paper.

Table1. The parameters (at 1.5GHz) of Co_2Z and Co_2Y at 1000°C

Sample	**ε′**	**tanδ$_ε$**	**μ′**	**tanδ$_μ$**
$3Ba_{0.7}Sr_{0.3}O \cdot 2CoO \cdot 10.8Fe_2O_3$	5.5	0.01	4.6	0.19
$Ba_2Co_2Fe_{12}O_{22}$	9.53	0.03	2	0.11

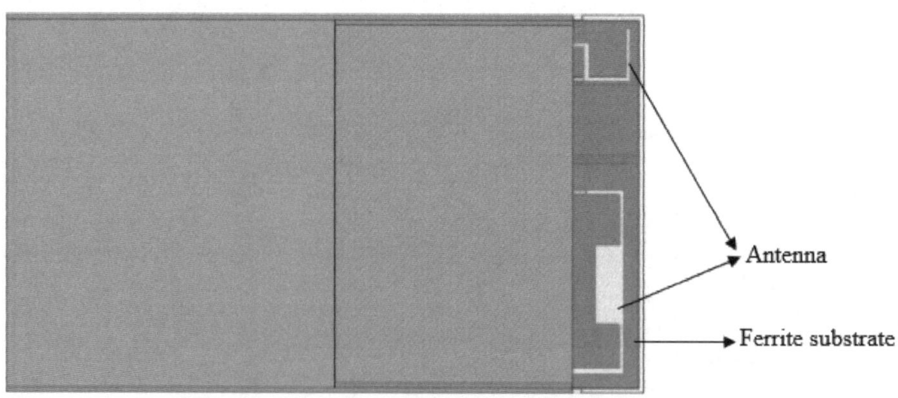

Figure3. The mobile phone antenna model.

IOP Conf. Series: Materials Science and Engineering **265** (2017) 012005 doi:10.1088/1757-899X/265/1/012005

Table1 is material parameters when we emulate by using HFSS. The focus frequency of antenna are all set at 1.5GHz. The antenna model is shown in Fig.3. Ferrite substrate covers the antenna radiation element. The S11 and RE of antenna can be seen from Fig.4, and Table2.

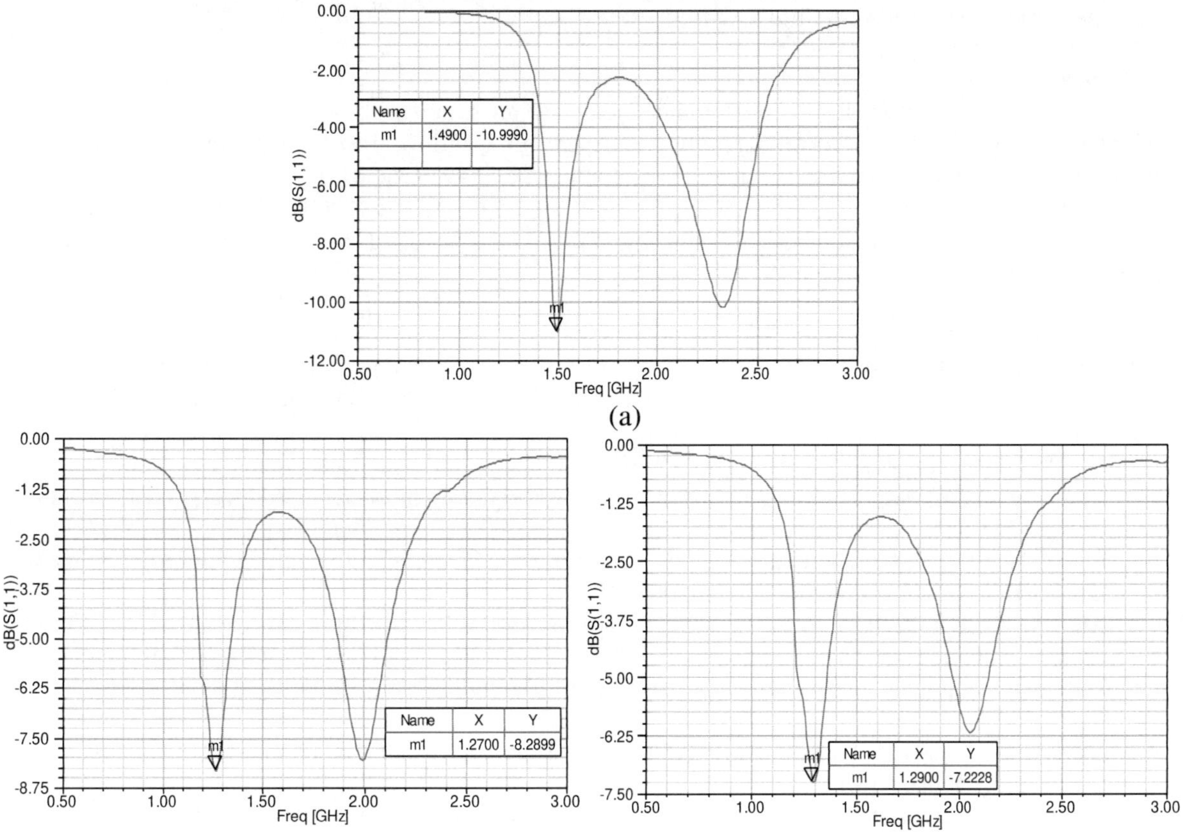

Figure4. The S11 parameter of antenna at 1.5GHz. (a) Without ferrite substrate; (b) Z-type ferrite substrate; (c) Y-type ferrite substrate.

Table2. The RE of different substrate antenna at 1.5GHz.

Substrate	**RE（radiation efficiency）**
None	82%
Z-type	64.37%
Y-type	68.39%

It can be seen when radiation efficiency and reflectivity of antenna is satisfied. Comparing with the antenna without ferrite substrate, the resonance frequency (at 1.5GHz) of antenna with ferrite substrate shift to lower frequency about 200MHz-220MHz, not only Z-type ferrite has good effect but Y-type ferrite affect well. From Table1, μ′ and Ɛ′ of $3Ba_{0.7}Sr_{0.3}O \cdot 2CoO \cdot 10.8Fe_2O_3$ ferrite can be realized nearly to match, therefore, electromagnetic wave (EM) can enter the antenna structure easier. On the other hand, the magnetic tangent loss and dielectric tangent loss are little relatively, the energy of EM can't lose much in the structure. Permittivity (ε′) of $Ba_2Co_2Fe_{12}O_{22}$ do not match well with μ′, but it has lower magnetic loss than $3Ba_{0.7}Sr_{0.3}O \cdot 2CoO \cdot 10.8Fe_2O_3$. It's all known that physical dimension of antenna decrease with increasing of frequency. Thus, it is possible to design a smaller size antenna to achieve miniaturization of a little low-frequency antenna.

4. Conclusion

To get high cut-off frequency of hexagonal ferrite, it can sinter at 1000°C with the aid of Bi_2O_3, permeability parameter and sintering temperature are in the direct ratio. As antenna is the most important part of mobile phone, to miniature antenna size, it is proposed to find the electromagnetic parameter of $3Ba_{0.7}Sr_{0.3}O \cdot 2CoO \cdot 10.8Fe_2O_3$ and $Ba_2Co_2Fe_{12}O_{22}$ at 1000°C to apply in magnetic substrate. At 1.5GHz, the resonance frequency 200MHz-220MHz of antenna can move 200MHz-220MHz, and RE diminish not much.

References

[1] Zhang H, Li L, Zhou J, et al. Microstructure characterization and properties of chemically synthesized Co 2 Z hexa-ferrite[J]. Journal of the European Ceramic Society, 2001, 21(2): 149-153.

[2] KONG Zhan-xing，LI ANG Di-fei, XI E Jian-1iang，DENG Long-jiang. Effect of Bi 203 Adding and Sintering Temperature on the Cut-off Frequency of Ni Zn Ferrites [J]. J Magn Mater Devices,2008,(06):61-64

[3] Bae S, Hong Y K, Lee J J, et al. Co_2Z hexa-ferrite T-DMB antenna for mobile phone applications [J]. IEEE Transactions on magnetics, 2009, 45(10): 4199-4202.

[4] Xiang C C, Nie Y, Feng Z K. Low-loss Z-type hexa-ferrite with Sr substitution for microwave antenna miniaturization[C]//Microwave, Antenna, Propagation, and EMC Technologies for Wireless Communications (MAPE), 2011 IEEE 4th International Symposium on. IEEE, 2011: 71-74.

[5] Qi Xia. Low-loss magneto-dielectric materials and the applications in antennas [D]. University of Electronic Science and Technology of China, 2013.

[6] Chang ling. Research and Design on Novel Antenna for Mobile Phone Application [D]. XIDIAN UNIVERSITY, 2014.

A metal-free route to synthesize pure-phase 3C-SiC with excellent optical and magnetic properties

S Liu[1], J G Wang[1,2,*]

[1]Jiangsu Key Laboratory of Advanced Metallic Materials, School of Materials Science and Engineering, Southeast University, Nanjing 211100, P.R. China
[2]Xizang Key Laboratory of Optical Information Processing and Visualization Technology, School of Information Engineering, Xizang Minzu University, Xianyang 712082, P.R. China.

Email addresses: wangjigang@seu.edu.cn; fred2015123@sina.com

Abstract. Large amount of pure-phase 3C-SiC particles were synthesized via microwave heating, without using any metal catalyst. Ball-milling pre-treatment was employed to enhance reaction activity of raw materials prior to the microwave heating process. The morphology, microstructure of the SiC products were characterized by scanning electron microscope (SEM), X-ray diffraction (XRD), and transmission electron microscope (TEM). As the result of SEM, the SiC particles have diameters of 500–2000 nm and smooth surface. TEM image shows that some of the micro-sized SiC particles are composed of agglomerate nano-particles with diameters of 50–200 nm. Photoluminescence and magnetic properties of the SiC products were measured by fluorescence spectrophotometer and vibrating sample magnetometer (VSM), respectively. Ultra-violet emission from the 3C-SiC products can be detected under excitation wavelength of 240 nm. Without any contribution from the magnetic metal ions, the SiC particles can exhibit ferromagnetic properties around room temperature with saturation magnetization (Ms) of 0.9 emu/g, approximately. The excellent optical and magnetic properties of SiC may mainly be attributed to the point defects.

1. Introduction

In current years, compound semiconductors such as ZnO, GaN, and SiC, etc. have draw considerable attentions because their promising applications in optical and electron devices. Many research works have been carried out to synthesize these semiconductors with excellent optical and electrical properties [1-3]. Compared with other compound semiconductors, SiC may be utilized in high-power and high-frequency devices because of the excellent fundamental properties covering high thermal conductivity, high thermal stability, and high mechanical strength [4-6]. Thus, numerous studies on the optical properties of SiC have been made for the practical application in various devices.

Photoluminescence properties were commonly measured for evaluating the luminous power of SiC products. Up to now, violet, blue emissions from 3C-SiC products have been found [7-9]. The detected emission peaks of these 3C-SiC products show the blue-shift phenomena compared with the intrinsic luminous peak of bulk 3C-SiC (Eg=2.23 eV). The blue-shift phenomena were always suggested to be attributed to quantum confinement effect [10], oxygen vacancies [11], internal or surface point defects [12] etc. However, the origin of the blue-shift phenomenon is still in debate, and metal impurities which can affect the photoluminescence properties were easily introduced into SiC products in many synthesis processes [13, 14]. Thus, exploitation of metal-free synthesis for preparing pure-phase SiC with excellent photoluminescence properties is still desirable.

Since the proposal of spintronic devices [15], potential application of semiconductors in magnetic devices and fundamental properties of SiC mentioned above lead to larges numbers of studies on the improvement or origin of the magnetic properties of SiC [16-19]. Transition metals including Mn, V, Fe etc. were always introduced into SiC for the preparation of ferromagnetic SiC. However, the precipitation of metal-related secondary phase may limit the understanding of the origin of ferromagnetism of SiC and the application of ferromagnetic SiC at high temperature. Thus, exploration a metal-free route to obtain ferromagnetic SiC have been focused on. The most common methods without any metal impurity is the introduction of Si- or C-related vacancies [20,21], but most of them need complex procedure or strict reaction conditions.

In our previous study [22, 23], we found that the SiC nanowires synthesized via microwave heating exhibit excellent photoluminescence properties or weak room-temperature ferromagnetism. The excellent optical and magnetic property were proposed to be attributed to the surface or internal defects in SiC products. The high reaction rate under microwave heating with high heating rate may benefit the formation of defects covering stacking faults or point defects, further result in excellent functional properties. However, the luminous efficiency and Ms of the SiC products is still need to be enhanced.

In this study, high-energy microwave heating method is also employed to synthesize SiC products because it is a simple and faster process [24]. No metal or other impurity was added in the raw materials. For preparing SiC products with better optical and magnetic properties, ball-milling pre-treatment was employed to enhance the reaction rate. Large amount of micro-sized pure-phase SiC with excellent photoluminescence and magnetic properties were prepared. And also, the origins of the optical and magnetic performance of the obtained SiC products are briefly discussed.

2. Experimental sections

2.1 Synthesis of 3C-SiC products

High-purity silicon (⩾99.99%), silica (⩾99.0%) and graphite powders (⩾99.85%) (mole ratio 1:1:2) were used as raw materials. The mixed powders were ball-milled by 500 agate grinding balls with rotation speed of 400 r/min for 6 h. The crucible together with the specimens was then placed at the center of a commercial microwave vacuum sintering furnace with the type of NJZ10-3 (Nanjing Jiequan Microwave Co. China). Prior to the heating process, the furnace cavity was evacuated below 10^4 Pa by a watering circulating pump. The employed microwave power was in the range of 3-5 kW. With the use of the microwave heating system, the temperature of the raw powders can be heated over 1500 ^{0}C in several minutes. The specimen was kept at the designed temperature (approximately 1350 ^{0}C) for 1 h. The temperature was measured by an optical pyrometer (Reytek) on the top of insulation glass. The temperature of the products can be rapidly decreased, and at last light-green powders can be obtained.

2.2 Characterizations

The phase structure of the product were examined by X-ray diffraction (XRD), which were carried out on a D8 ADVANCE X-ray diffractometer (Bruker Co., German) with Cu Kα radiation (λ=1.5418 Å). Morphologies of the product were examined by a field emission scanning electron microscopy (FEI Ltd., Netherlands). Further structural analysis of the products was performed using a FEI Tecnai T20 microscopy (FEI, Eindhoven, Netherlands).

The photoluminescence spectra were measured using a Hitachi F-7000 typed fluorescence spectrophotometer at room temperature, with the excitation light of 240 nm from Xe lamp. The magnetization versus magnetic field (M-H) curves were measured using LakeShore 7407 vibrating sample magnetometer (VSM) at room temperature. Data were collected over the magnetic field of -10000—10000 Oe (250/π A*m-1).

3. Results and discussion

3.1 Morphology, microstructure of the as-obtained SiC product

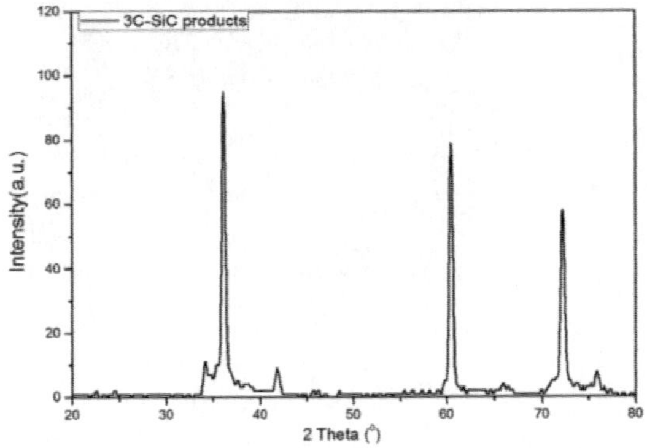

Figure 1. XRD pattern of the light-green products.

The XRD were employed for investigating the phase structure of SiC. In the XRD patterns (figure 1), the peaks at 35.6, 41.5, 60.0, 71.8 and 75.6 ° correspond to the (111), (200), (220), (311) and (222) of 3C-SiC (JCPDS No. 29-1129). From the XRD patterns of the products, it can be found that the are the product is mainly composed of 3C-SiC. According to the XRD results of our previous study [22], other impurities such as silica and graphite may retained in the SiC products after microwave heating process (1-2 h). Different from the previous results, no other impurity can be found in the XRD patterns, indicating that all the raw materials were completely converted into pure-phase 3C-SiC. The formation of pure-phase 3C-SiC suggest that ball-milling pre-treatment can significantly enlarge the reaction velocity of raw materials in the microwave heating process.

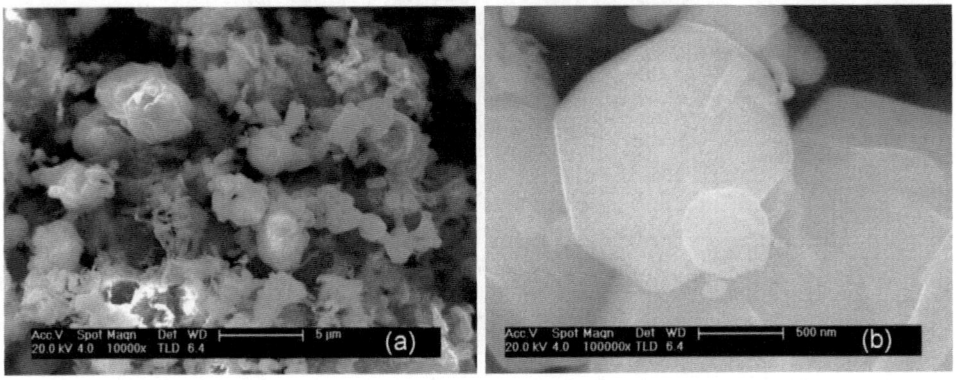

Figure 2. SEM images of the 3C-SiC products with different magnifications.

The morphology and the elemental composition of the SiC products were characterized by SEM and EDS, respectively. Large amount of nano- or micro-sized SiC particles can be found as shown in the SEM images (figure 2a), and the sizes of most SiC particles are in the range of 500-2000 nm. The SEM image with larger magnification (figure 2b) shows that the as-obtained SiC particles have clean and smooth surface. In the previous studies on the synthesis of nano- or micro-sized SiC [7, 10, 25], silica layers could always be found on the surface of SiC core. However, oxygen element can not be detected within the resolution of EDS (figure 3). There are three possible reasons for explaining the absence of oxygen elements: 1) Because of the high formation rate of the SiC, oxidizing atmosphere is rare in the oxidation process, and thus the SiC core can hardly be oxidized; 2) The formed silica layer can easily be reduced in the reducing atmosphere because of the high reaction activity of the ball-milled graphite powders; 3) The oxygen atoms on the surface are rather few, and thus the oxygen element can not be detected by EDS.

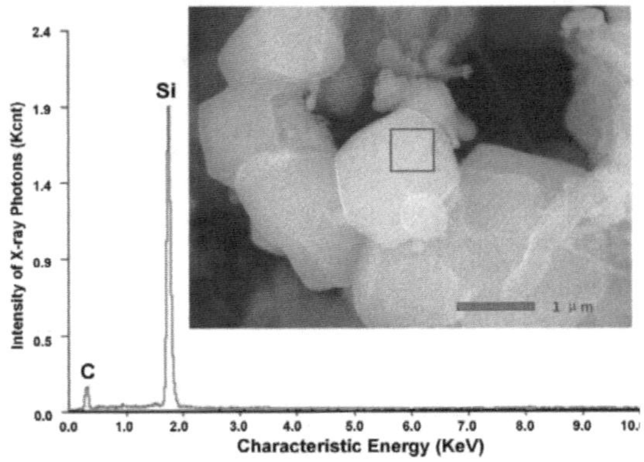

Figure 3. EDS spectrum of the as-obtained 3C-SiC particles.

TEM was employed for further information on the microstructures of the SiC particles. From the TEM images (figure 4), it can be found that the micro-sized SiC particles were composed of agglomerate SiC nano-particles. Furthermore, no silica layer can be detected by TEM, and the absence of the silica layer is consistent with the results of the EDS. Selected area electron diffraction (SAED) analyses (inset in figure 4) further confirm that the products are mainly composed of 3C-SiC. The inserted SAED pattern exhibits clear rings corresponding to 3C-SiC {111}, {220} and {311} planes.

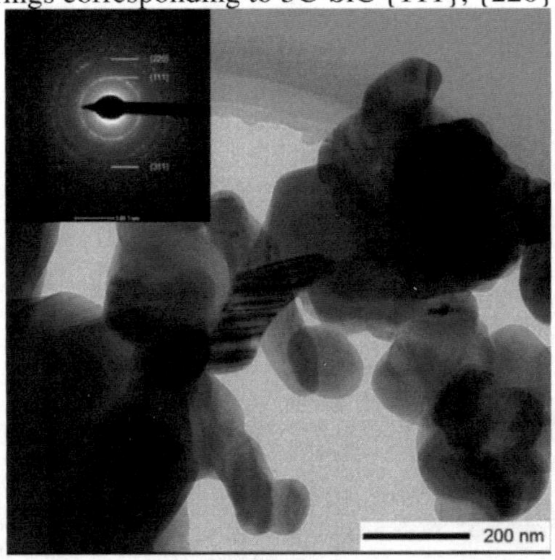

Figure 4. TEM image and SAED pattern (inset) of the as-obtained 3C-SiC particles.

3.2 Photoluminescence and magnetism investigations

In this study, ultra-violet emission peak centered at 390 nm are detected under excitation wavelength of 240 nm (figure 5). The broad peak shows a larger emission wavelength compared with the peak under excitation wavelength of 325 nm [22]. It is noteworthy that the broad peak covered the wavelengths corresponding to the E_g of 4H-SiC and 6H-SiC. Besides, the similar ultra-violet emission was also found from 6H-SiC nanowires [7]. Thus, the ultra-violet emission was initially thought to be caused by various polytypes of SiC in the products. However, based on the XRD patterns of the products, the main consistent of the product is 3C-SiC. And also, as the large size of the SiC particles, the size effect or quantum confinement effect can not contribute to the ultra-violet emission from the SiC products.

IOP Conf. Series: Materials Science and Engineering **265** (2017) 012006 doi:10.1088/1757-899X/265/1/012006

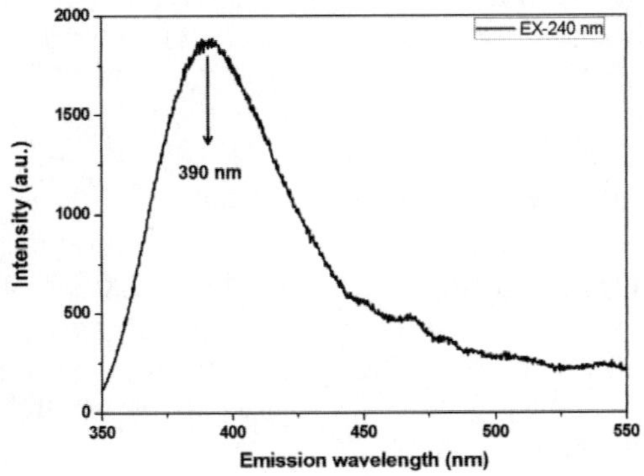

Figure 5. Photoluminescence spectrum of the as-obtained 3C-SiC particles.

The oxygen vacancies are also employed to explain the photoluminescence properties of oxide, such as nano-sized SiO_2 [26] or ZnO [27]. In these explanations, the oxygen vacancies can provide a site for electron-hole recombination which can cause the blue-shift phenomena compared with well-crystallized bulk materials. However, the oxygen vacancies in the silica layer should not be the single origin of the ultra-violet emission because rare oxygen can be detected by EDS and TEM. Similar to the role of oxygen vacancies in the ultra-violet emission, there may be other point defects which can also provide sites for the electron-hole recombination, such as Si-related or C-related vacancies.

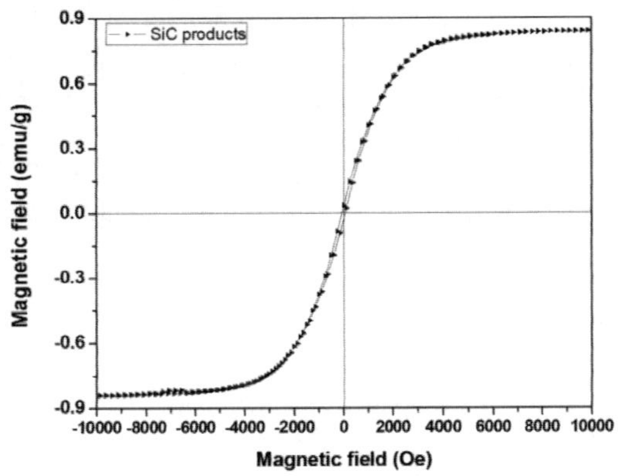

Figure 6. M-H curve of the as-obtained 3C-SiC particles.

The M-H curve (figure 6) of the obtained 3C-SiC particles exhibit ferromagnetism around room temperature with Ms of 0.9 emu/g and coercivity (Hc) of 70 Oe, approximately. In previous studies, the most effective method to prepare ferromagnetic SiC was reported to be doping transition metal into SiC. For example, the Ms can be enhanced to 0.06, 0.002 and 1.3 emu/g by doping small quantities of Mn, Cr, and Fe, respectively [18, 28, 29]. In our experiment, the magnetic moments of SiC can be significantly enhanced via employing ball-milling pre-treatment and microwave heating method without any metal impurity. It is worth noting that the Ms of the as-prepared SiC is close to the Ni-catalyzed SiC products in our previous study [23]. The ferromagnetic response cannot be attributed to any other impurities since the as-employed raw materials covering silicon, silica and graphite powders did not exhibit ferromagnetism signals.

The origin of ferromagnetic signals from semiconductors without metal impurities is still in debate. Among explanations of the magnetic exhibitions of SiC, the most acceptable contributions are the disorders and point defects covering Si vacancy, C vacancy, and Si-C bi-vacancies in SiC [19-21]. In our previous study [23], high reaction rate under microwave irradiation benefits the formation of point defects which can result in the ferromagnetism exhibition of SiC. And also, ball-milling pre-treatment which can further enhance the growth rate of SiC can be employed to enhancing the Ms of SiC. Thus, the ferromagnetic signals from the as-obtained 3C-SiC products can be attributed to the internal point defects of SiC. The present work develop a new approach for preparing ferromagnetic SiC without using any transition metals.

4. Conclusions

Pure-phase 3C-SiC can be synthesized by employing ball-milling pre-treatment and microwave heating. The morphology of the SiC products is relative uniform, and the SiC particles have clean and smooth surface. Some of the micro-sized SiC with size of 500-2000 nm are composed of agglomerate SiC nano-particles with size of 50-200 nm. These micro-sized SiC particles can emit ultra-violet light under excitation wavelength of 240 nm. The ultra-violet light may be attributed to the point defects. Without any ferromagnetic metal impurity, the as-obtained SiC exhibit room-temperature ferromagnetism with Ms of 0.9 emu/g and Hc of 70 Oe. The ferromagnetic performance of SiC particles may also mainly be contributed by the internal or surface point defects of SiC.

Acknowledgements

This work is supported by Program for New Century Excellent Talents in University (NECT-12-0119), Key Project of Science and Technology of Tibet Autonomous region (2015ZR-14-14), Qing-lan Project of Jiangsu Province, Summit of the Six Top Talents Program of Jiangsu Province (2013-JY-007), and the Fundamental Research Funds for the Central Universities.

References

[1] Li Z Q, Zhang D X and Lin J J 2006 *J. Appl. Phys.* **99** 124906
[2] Koblmüller G, Reurings F, Tuomisto F and Speck J S 2010 *Appl. Phys. Lett.* **97** 191915
[3] Seong H, Choi H, Lee S, Lee J and Choi D 2004 *Appl. Phys. Lett.* **85** 1256-58
[4] Shim H W, Kuppers J D and Huang H C 2008 *J. Nanosci. Nanotechno.* **8** 3999-4002
[5] Turner S P, Taylor R, Gordon F H and Clyne T W 1993 *J. Mater. Sci.* **28** 3969-76
[6] Liu H T, Cheng H F, Wang J and Tang G P 2010 *Ceram. Int.* **36** 2033-37
[7] Wei G D, Qin W, Wang G, Sun J, Lin J, Kim R, Zhang D and Zheng K 2008 *J. Phys. D: Appl. Phys.* **41** 235102
[8] Zhu J, Wu H, Chen H T, Wu X L and Xiong X 2009 *Phys. Lett. A* **373** 1697-1700
[9] Zhang J, Yan S, Jia Q, Huang J, Lin L and Zhang S 2016 *Physica E* **80** 19-24
[10] Wu X L, Fan J Y, Qiu T, Yang X, Siu G G and Chu P K 2005 *Phys. Rev. Lett.* **94** 026102
[11] Li Z, Zhao J, Zheng M, Xia J and Meng A 2014 *Nano Res.* **7** 462-72
[12] Liu X and Yao K 2005 *Nanotechnology* **16** 2932-35
[13] Xi C, Peng Y, Wan S, Li T, Yu W and Qian Y 2004 *J. Phys. Chem. B* **104** 20102-04
[14] Wang F, Qin X, Zhu D, Meng Y, Yang L, Sun L and Ming Y 2015 *Mater. Sci. Semicond. Process.* **29** 155-160
[15] Prinz G A 1998 *Sci.* **282** 1660-63
[16] Yu L, Jin H, Liu D, Dai Y, Guo M, Huang B and Zhang Z 2010 *Chem. Phys. Lett.* **496** 276-79
[17] Yan Y L, Men B Q, Liu G S and Zheng H W 2014 *Appl. Phys.* **117** 841-45
[18] Zheng H W, Wang Z Q, Liu X Y, Diao C L, Zhang H R and Gu Y Z 2011 *Appl. Phys. Lett.* **99** 222512
[19] Seong H, Park T, Lee S, Lee K, Park J and Choi H 2009 *Met. Mater. Int.* **15** 107-11
[20] Liu Y, Wang G, Wang S, Yang J, Chen L, Qin X, Song B, Wang B and Chen X 2011 *Phys. Rev. Lett.* **106** 087205
[21] Li L, Prucnal S, Yao S D, Potzger K and Anwand W 2011 *Appl. Phys. Lett.* **98** 222508
[22] Wang J, Liu S, Ding T, Huang S and Qian C 2012 *Mater. Chem. Phys.* **135** 1005-11
[23] Liu S and Wang J 2016 P*hysica E: Low-dimensional Systems and Nanostructures* **81** 268-74
[24] Nüchter M, Ondruschka B, Bonrath W and Gum A 2004 *Green Chem.* **6** 128-41
[25] Liu X, Cao L, Song H and Jiang H 2014 *Physica E* **61** 167-70
[26] Yu D P, Hang Q L, Ding Y, Zhang H Z, Bai Z G, Wang J J, Zou Y H, Qian W, Xiong G C and Feng S Q 1998 *Appl. Phys. Lett.* **73** 3076-78
[27] Singh J, Kumar P, Hui K S, Hui K N, Raman K, Tiwari R S and Srivastava O N 2012

CrystEngComm **14** 5898-5904
[28] Huang Z and Chen Q 2007 *J. Magn. Magn. Mater.* **313** 111-14
[29] Song B, Jian J K, Li H, Lei M, Bao H Q, Chen X L and Wang G 2008 *Physica B* **403** 2897-2901

Effect of Fe_2O_3 addition on microstructure and dielectric properties of PbO-SrO-Na_2O-Nb_2O_5-SiO_2 glass ceramic

F H Tan[1], Q M Zhang[2], H B Zhao[1], F Wei[1], P Zhou[2], J Y Chen[2] and J Du[1,3]

[1] State Key Laboratory of Advanced Materials for Smart Sensing, General Research Institute for Nonferrous Metals, Beijing 100088, China.
[2] Advanced Electronic Materials Institute, General Research Institute for Nonferrous Metals, Beijing 100088, China.
[3] e-mail: dujun@grinm.com.

Abstract. Glass ceramic with the addition of 0-2.0 mol% Fe_2O_3 were synthesized by cast quenching followed by controlled crystallization, and the effect of Fe_2O_3 addition on the microstructure and dielectric properties of PbO-SrO-Na_2O-Nb_2O_5-SiO_2 (PSNNS) glass ceramic were investigated. DTA measurements revealed that the crystallization temperature of $(Pb,Sr)Nb_2O_6$ can be decreased by adding Fe_2O_3, meanwhile, the grain sizes of PSNNS glass ceramic increase with the increasing amount of additives. Dielectric properties of PSNNS glass ceramic were obviously affected by Fe_2O_3 addition. With 1.5 mol% Fe_2O_3 addition, the dielectric constant increased from 720 to 917, the dielectric loss increased from 0.012 to 0.0124. The PSNNS glass ceramic with 1.5% Fe_2O_3 addition show great promise for high voltage capacitors applications.

1. Introduction

Niobate glass ceramic used as dielectric material for high voltage capacitors and high energy storage capacitor has attracted increasing attention in recent decades [1-6]. The excellent dielectric properties for glass ceramic are mainly due to their unique microstructure. Glass ceramic, which are composites of nanocrystallites embedded in the glass matrix obtained by reheating glasses at high temperatures. Ceramic phase have large permittivity, glass matrix enjoys high electric breakdown field strength [7]. Meanwhile, the dielectric properties of niobate glass ceramic were greatly affected by crystal phase as well as glass phase. Since the glass ceramic have high breakdown strength(>90 kV/mm) [8], the key to improving the dielectric performance of glass ceramics is to increase the dielectric constant, numerous methods was investigated, such as change the ratio of Pb and Sr [9], rare earth elements addition [10] and titanate glass ceramics [11].

Fe_2O_3 is probably one of the most frequency used acceptors in ferroelectrics, the effects of Fe_2O_3 on titanate ceramics and lead zirconate titanate ceramics have been studied [12-15]. The found that dielectric properties could be enhanced by substituting the ions at the A or B site of the ABO_3 perovskite structure [16]. However, information regarding the effect of Fe_2O_3 addition on the glass ceramic composites materials was scarcely reported. The purpose of this study was to investigate the effect of Fe_2O_3 modification on the microstructure and dielectric properties of PbO-SrO-Na_2O-Nb_2O_5-SiO_2 (PSNNS) glass ceramic. Based on the previous research, $P_{0.6}S_{0.4}$NNS was chosen as the base material system [9], and then Fe_2O_3 was added according to different ion ratios.

2. Experimental

Reagent grade PbO, SrO, Na_2O, Nb_2O_5, SiO_2, Fe_2O_3 were used as starting materials. The basic composition of $P_{0.6}S_{0.4}$NNS powder was first prepared and then added with x mol% Fe_2O_3 (where x =0,

0.1, 0.5, 1, 1.5, 2). The samples with increasing Fe_2O_3 concentration were denoted as PSNNS-0, PSNNS-0.1, PSNNS-0.5, PSNNS-1, PSNNS-1.5 and PSNNS-2. Mixtures of the desired composition were ball-milled with zirconia balls as grinding media and deionized water as a lubricant for 12 h. After mixing and milling, the powders were dried for 12 h. Then, well-mixed oxide powders containing appropriate amounts of the above compositions were melted in a platinum crucible for 2-3 h at 1420 °C. The initial glasses were formed by rapid melt-quenching. The temperature of controlled crystallization of glass sheets was determined by Differential Thermal Analysis (DTA; Model WCT-2C, BoFei, Beijing) analysis method. After the crystallization temperature was determined, the initial glass bulks gone through the controlled crystallization process. Then the glass ceramics were cut and polished into sheets of 4 mm in thickness. Gold electrodes were sputter deposited on both sides of the glass ceramic sheets.
The measurements of dielectric constant and dielectric loss for PSNNS glass ceramics were performed using 4284A precision LCR meter. And the crystal structure of the samples was analyzed using X-ray diffractometry (XRD; Model MSAL-XD2, Micro Structure Analysis Laboratory, Beijing) with a 2θ range from 10°-90°. Microstructure evolution was observed using scanning electron microscopy (SEM; Model Hitachi S4800) equipped with a field emission gun. Dielectric properties including leakage current and polarization-electric field (P-E) were measured by ferroelectric tester (RT6000HVA, Radiant Technology, Albuquerque, NM).

3. Results and discussion

Figure 1 demonstrates the DTA curves of PSNNS-0, PSNNS-0.1, PSNNS-0.5, PSNNS-1, PSNNS-1.5 and PSNNS-2. The inflection around 630~670 °C on the DTA curves was considered as the glass transition endothermic effect. Three main exothermal crystallization peaks were observed on the curve of each sample. It's obvious that the crystallization temperature of PSNNS glass could be decreased by adding Fe_2O_3. It's well known that Fe_2O_3 could acted as a nucleating agent during the crystallization process, thus, ferroelectric phases were more easily crystallized from the glass matrix, PSNNS crystallization process was promoted. On the previous research[9], the crystallization process of the PSNNS glass was studied in detail, on this basis the crystallization temperature was chosen at 1000 °C for 3 h in this study, and the XRD patterns of the samples were shown in Fig. 2 to see if there was phases change when Fe_2O_3 was added to the precursor.

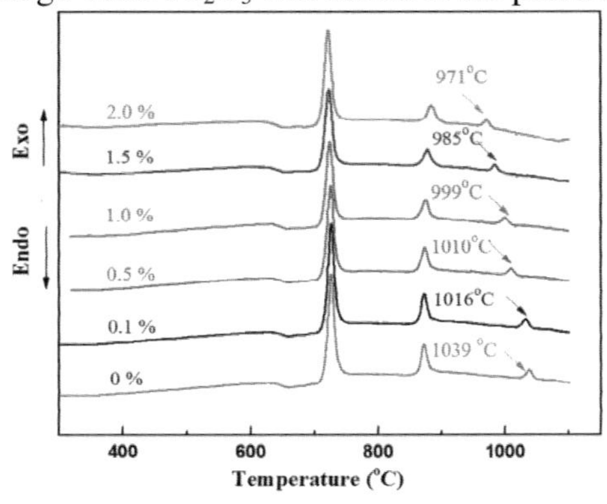

Figure 1. DTA curve of Fe_2O_3 doped PSNNS glass ceramic.

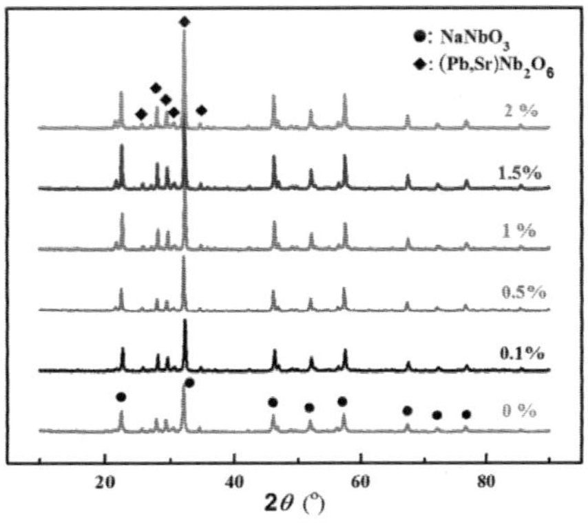

Figure 2. XRD patterns of PSNNS glass ceramic with different Fe_2O_3 addition.

The XRD patterns of PSNNS glass ceramic with the addition of 0-2 mol% Fe_2O_3 are shown in Fig. 2; all of the samples were crystallized at 1000 °C for 3 h. As is shown, the main crystal phases of PSNNS glass ceramic is $(Pb,Sr)Nb_2O_6$ with tungsten bronze structure and $NaNbO_3$ with perovskite structure. The dielectric constant of $(Pb,Sr)Nb_2O_6$ and $NaNbO_3$ is about 200 and 600 [17,18], respectively, They are responsible for raising the overall dielectric constant of PSNNS glass ceramic. Compared with the results of DTA and XRD, it can be concluded that there was no phase changed with Fe_2O_3 addition. Moreover, with the addition of Fe_2O_3, it exhibits a more sharp diffraction peak and the diffraction intensity was enhanced, indicted that PSNNS glass ceramic with higher degree of crystallization was obtained though Fe_2O_3 addition.

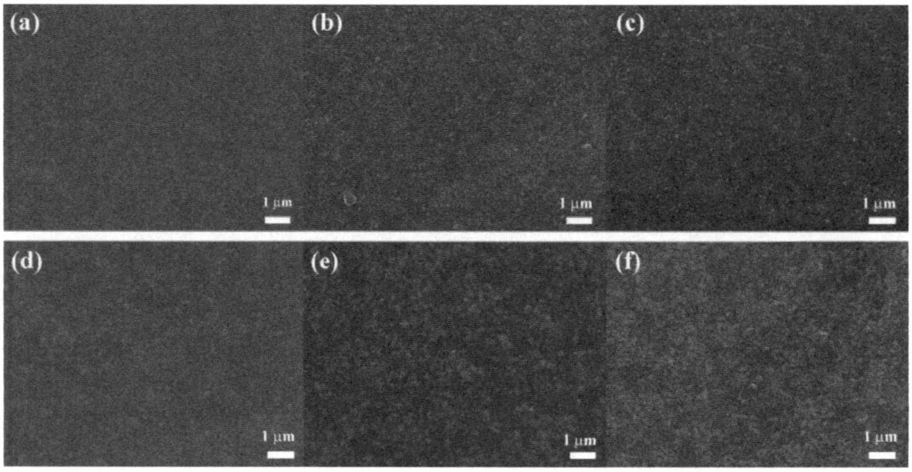

Figure 3. SEM photographs of the PSNNS specimens with different Fe_2O_3 addition: (a) 0.0 mol%, (b) 0.1 mol%, (c) 0.5 mol%, (d) 1.0 mol%, (e) 1.5 mol%, (f) 2.0 mol%.

Figure 3 shows the SEM photographs of the PSNNS glass ceramic with different Fe_2O_3 addition crystallized at 1000 °C. Seen from the image of the microstructure, the PSNNS glass ceramic was dense, smooth and without holes and cracks. The crystal grains are uniformly dispersed, suggesting that the rate of nucleation is relatively high, and the transformation of glass phase to crystalline phase is therefore completed with a minimum of time. Furthermore, Fe^{3+} in glass matrix acted as a nucleating agent during crystallization process, therefore, there are more ceramic phase precipitated from the glass matrix. It is seen from the photograph that the grain sizes remarkably increase with the increasing of Fe_2O_3 addition. In general, the increase in grain size results in an increase of dielectric constant [19], furthermore, the addition of Fe_2O_3 can reduce the crystallization activation energy and promote the crystallization of ceramic phase, and ultimately improved the overall dielectric constant.

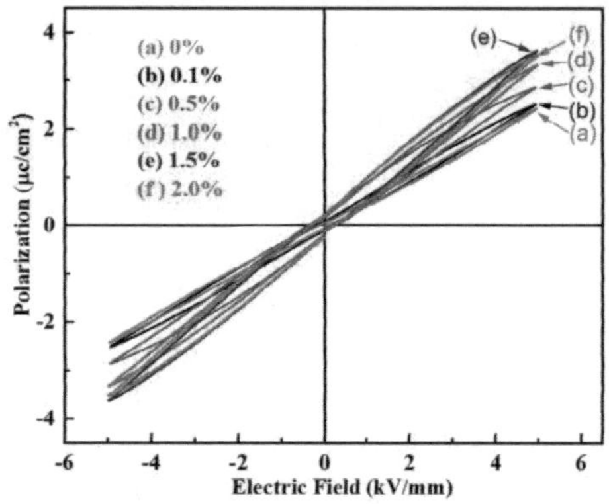

Figure 4. Polarization-electric field loops of the PSNNS specimens with different Fe_2O_3 addition.

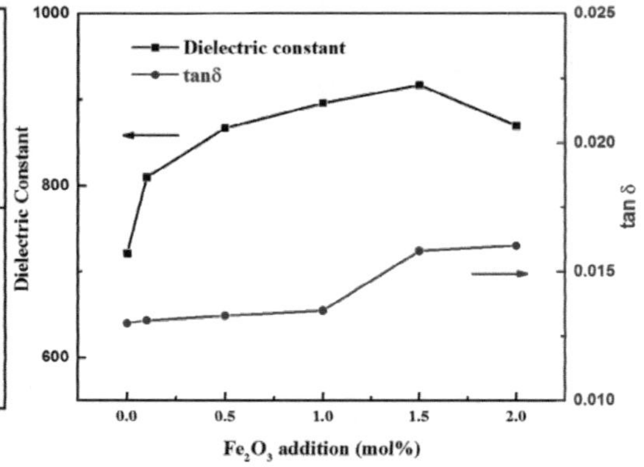

Figure 5. Dielectric constant and loss tangent of PSNNS glass ceramic with different Fe_2O_3 addition.

The polarization-electric field ferroelectric hysteresis loops were carried out at room temperature for the glass ceramic samples in order to verity the effect of Fe_2O_3 addition on the polarization characteristics in Fig. 4. The plots of hysteresis loops driven at the same electric of 5 kV/mm for all samples. Obviously, with the increment of the concentration of Fe_2O_3, the P-E hysteresis loops of the glass-ceramic samples grow thicker, and the coercive field (E_c), the maximum polarization (P_{max}), and remnant polarization (P_r) gradually increase with the increasing concentration of Fe_2O_3 additive. All the hysteresis loops are of the nature of non-saturation, which is due to the small grain size of the crystal particles precipitated within the glass matrix and the stresses applied by the surrounding glass matrix [20].

All of the dielectric property was measured at room temperature by using an Agilent 4284A precision LCR meter. Fig. 5 shows the plots of the dielectric constant and dielectric loss of the PSNNS glass ceramic with x mol% Fe_2O_3 addition ($0 \leq x \leq 2$). The measurement result demonstrates that the dielectric constant increases with the increasing amount of Fe_2O_3 addition in the range of $0 \leq x \leq 1.5$, whereas it decreases slightly in the range of $1.5 \leq x \leq 2$. A maximum dielectric constant of 917 was obtained at compositions with 1.5 mol% Fe_2O_3 additions. Furthermore, it can be seen that the dielectric loss increased slightly when the Fe_2O_3 content is increased from 0-1 mol%, the increase in dielectric loss is mainly due to the accumulation of space charge at grain boundary phases. Then the dielectric loss increased significantly when the addition of Fe_2O_3 exceeds 1 mol%, probably due to the existence of metastable Fe^{3+} ions. Acceptor dopants Fe^{3+} create oxygen vacancies for ionic charge compensation and tend to pin the domain wall, which is evident by an increase in the internal bias field. Oxygen vacancies acted as carriers produce electrical losses.[11] Moreover, Fe^{3+} ions may trapped by defects or the interfaces in the dielectric materials, interfacial polarization occurs at the interface between the glass phase and the ceramic phase, polarization relaxation of the interfacial polarization is detrimental to the dielectric loss.

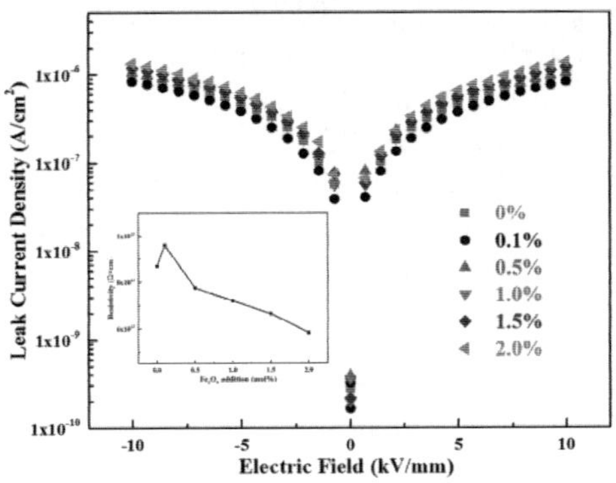

Figure 6. Leakage current density of PSNNS glass ceramic with different Fe_2O_3 addition.(Inset is the resistivity of PSNNS specimens)

The relationship of leakage current density and applied electric field for the PSNNS glass ceramic is given in Fig. 6. As usual, the leakage current increase with the applied field. It is seen that at the electric field of 10 kV/mm, the sample with 0.1 mol% Fe_2O_3 additions shows the minimal leakage current density of 9×10^{-7} A/cm^2, while the maximum leakage current density of 1.3×10^{-6} A/cm^2 was obtained with the composition of 2 mol% Fe_2O_3 additions. For dielectric materials, the low dielectric loss is often indicative of high resistivity, which can be confirmed by the resistivity measurements results of Fe_2O_3 doped PSNNS glass ceramic, and the results is given in the inset in Fig.6. When 0.1 mol% Fe_2O_3 was added to the glass, PSNNS-0.1 had the highest resistivity, when the mount of exceeded 0.1 mol%, the resistivity decreased, which matched the dielectric loss and leakage current well. And it confirmed that the Fe^{3+} occupy the substitution in the lattice when x is less than 1 mol%, while larger x would make the Fe^{3+} occupy the interstitial sites, thus the dielectric loss has an increment and resistivity decreases.

4. Conclusions

The effect of Fe_2O_3 doping on the microstructure and dielectric performance of PSNNS glass ceramic was investigated. DTA measurements revealed that the addition of Fe_2O_3 can decrease the crystallization temperature of $(Pb,Sr)Nb_2O_6$, SEM and XRD results suggest that the addition of Fe_2O_3 could refine the grains and enhance the degree of crystallization. With 1.5 mol% Fe_2O_3 addition, the dielectric constant increased from 720 to 917, the dielectric loss increased from 0.012 to 0.0124. However, excessive Fe_2O_3 is detrimental to the dielectric loss and leakage current. These excellent dielectric properties of PSNNS+1.5% glass ceramic indicate its promising application in electronic devices.

Acknowledgments
This work was supported by the National Natural Science Foundation of China (No.51477012).

References

[1] Reece M J, Worrell C A, Hill G J, Morrell R 1996 *J. Am. Ceram. Soc.* **79** 17
[2] Du J, Jones B, Lanagan M 2005 *Mater. Lett.* **59** 2821
[3] Zhou H, Zhang Q M, Tang Q, Cui H, Xu Y H, Lin C G, Du J 2015 *J. Electron. Mater.* **44** 3220
[4] Xiao S, Xiu S M, Xue S X, Shen B, Zhai J W 2016 *J. Mater. Sci.* **51** 5880
[5] Yang Y, Song J, Chen G H, Yuan C L, Li X Q, Zhou C R 2015 *J. Non-Cryst. Solids* **410** 96
[6] Zhou H, Zhang Q M, Tang Q, Cui H, Xu Y H, Lin C G, Du J 2015 *J. Electron. Mater.* **44** 4053
[7] Han D F, Zhang Q M, Luo J, Tang Q, Du J 2012 *Solid State Sci.* **14** 661
[8] Luo J, Du J, Tang Q, Mao C H 2008 *IEEE Trans. Electron. Devices* **55** 3549
[9] Han D F, Zhang Q M, Luo J, Tang Q, Du J 2012 *Ceram. Int.* **38** 6903
[10] Zhou Y, Zhang Q M, Luo J, Tang Q, Du J 2012 *Rare Metals*, **31** 281
[11] Chen M Y, Juuti J, His C S, Chia C T, Jantunen H 2015 *J. Eur. Ceram. Soc.* **35** 139
[12] Kim D H, Lee B W 2016 *J. Korean Phys. Soc.* **68** 574
[13] Chen C Y, Lin H L 2004 *Ceram. Int.* **30** 2075
[14] Yoon S J, Joshi A, Uchino K 1997 *J. Am. Ceram.* Soc. **80** 1035
[15] Zhu M K, Lu P X, Hou Y D, Wang H, Yan H J 2005 *J. Mater. Res.* **20** 2670
[16] Berlincourt D 1992 *J. Acoust. Soc. Am.* **91** 3034
[17] Reece M J, Worrell C A, Hill G J, Morrell R 1996 *J. Am. Ceram. Soc.* **79** 17
[18] Konieczny K, Kajtoch C 1998 *Ferroelectr.* **215** 65
[19] Wang X H, Chen R Z, Gui Z L, Li L T 2003 *Mater. Sci. Engin.* **99** 199
[20] Shyu J J, Chen C H 2003 *Ceram. Int.* **29** 447

IOP Conf. Series: Materials Science and Engineering **265** (2017) 012008 doi:10.1088/1757-899X/265/1/012008

Facile Fabrication of Highly Stretchable Nanocrack Silver Film using Magnetron Sputtering

Cancan Yang, Mei Yu, Chong Wang*, Zhe Yu.

Biomedical Microdevices Research Laboratory, Shenzhen Institutes of Advanced Technology, Chinese Academy of Sciences, 1068 Xueyuan Avenue, Shenzhen 518055, China.

* Corresponding Author e-mail: wangchong@siat.ac.cn

Abstract. Recently, stretchable electronic devices have been growing rapidly, such as bioelectrical interfaces, wearable and implantable electronics. Apparently, their stretchability highly depends on the surface structure. In this paper, highly stretchable nanocrack silver films were deposited using magnetron sputtering, which can be stretched by a maximum 150% strain while maintaining great electrical conductivity. Experimental results show that, compared to popular gold films, silver has relatively high electrical conductivity and better stretchability.

1. Introduction

Stretchable electronics[1-3]have great potential for serving as bioelectrical interfaces due to their better deformability[4] and biological compatibility[5]. Because of their outstanding deformability, the stretchable devices can also widely used as electronic skin[6,7], artificial prosthesis[8,9] and artificial retina[10]. Since the flexible micro-electrode array[11,12] first reported by Columbia[13] and Princeton University, stretchable electrodes have attracted significant interest and provides a promising method for wearable and implantable electronics.

To achieve stretchability, metal electrodes with nanocracks are fabricated on flexible substrates such as poly (dimethylsiloxane) (PDMS)[14,15]. Apparently, the surface structure, nanocracks, play an important role in the stretchability of metal films, which keep the electrodes conducting while being stretched at large strains. Previous work revealed that the nanocrack gold films could be stretched by a maximum 120% strain while maintaining their electrical conductivity [16]. The stretchable electronics are usually made from gold. Nevertheless, the electrical conductivity of gold is relatively low compared to other metals such as silver, cooper. Furthermore, as a kind of noble metal, gold is exceedingly expensive and thus the application of gold electrodes is extremely restrained. During the past decade, silver electrodes have been researched intensively due to its great electrical conductivity. However, up to now, the work of stretchable silver films are less reported.

Traditional electrodes are made usually by electroplate process[17,18], it plays an important role in electrical connection of sensitive electronic components. However, electroplate is a complicated and high-cost process to produce the electrodes. In this paper, we adopted a facile method to fabricate the stretchable nanocrack silver films using magnetron sputtering. Results illustrate that the silver films can be stretched by a maximum 150% strain while maintaining great electrical conductivity. In addition, compared to gold films, silver films have relatively high and better stretchability. Moreover, because of the lower cost, stretchable nanocrack silver film can be widely used in wearable devices.

2. Experiments

Figure 1 shows the process sequence for fabricating stretchable nanocrack silver films schematically, used as a rigid backing during fabrication, a silicon wafer (φ75mm) was first baked for 5 min at 200 ℃ to remove moisture and coated with a monolayer of 1H,1H,2H,2H perfilorooctyl-trichlorsilane (48931-10G Sigma Aldrich)to facilitate the removal of the PDMS membrane at the end of fabrication (figure 1(a)). The PDMS was mixed from the pre-polymer gel and the cross-linker (Dow Corning Sylgard 184) in a 10:1 (w / w) ratio, degassed in vacuum, spun on the silicon wafer at 600rpm for 60s, and then cured in an oven at 80℃ for at least 3 hours in order to obtain a 110μm thick PDMS membrane (figure 1(b)). To improve adhesion, the PDMS surface was first oxidized in an oxygen plasma for 30w, 30s (figure 1(c)). Patterned by a 200um thick stainless steel mask layer (figure 1(d) and (e)), a 20nm thick silver film was deposited (DC power 60W, 20s) on 4nm thick titanium buffer layer (DC power 40W, 5s). The fabrication process ended with manually peeling the PDMS membrane from Si wafer (figure 1(f)). On each Si wafer, several stretchable silver films are fabricated at the same time. They are subsequently cut with a pair of scissors into individual devices for electromechanical characterization. The stretchable silver film were similar to a dumbbell: the wire was designed to be 8mm long and 0.5mm wide, and at both the ends of the wire, there were 2 pads squares (1.5 mm × 1.5 mm), using to test the electrical resistances of the stretchable silver film.

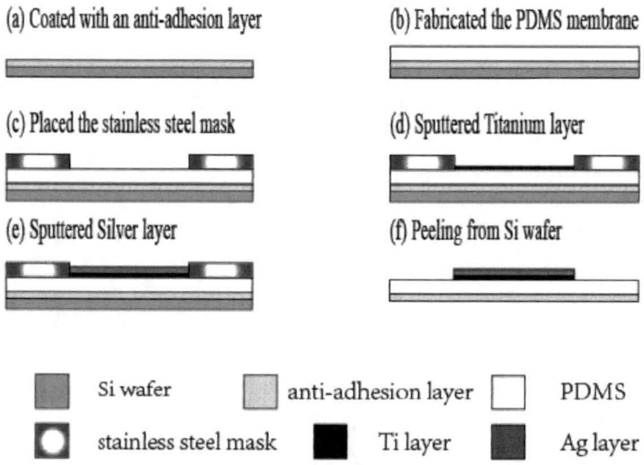

Figure 1. the processing sequence for fabricating the stretchable silver film.

3. Results and discussion

3.1 Crystalline structure and morphology

Figure 2(a) illustrates the typical XRD patterns of silver films. Diffraction peaks observed at 2θ= 38.1°, 44.2°, correspond well with (1 1 1), (2 0 0) 3C-syn silver phase (JCPDS No. 04-0783). Since no other impurity diffraction peaks are present, no other image is observed in the diffraction pattern, all diffraction peaks in the XRD pattern can be indexed to a silver with lattice constants of a=4.08 Å. The thickness of stretchable silver film is about 20nm, measured by AFM, demonstrated in Figure 2(b).

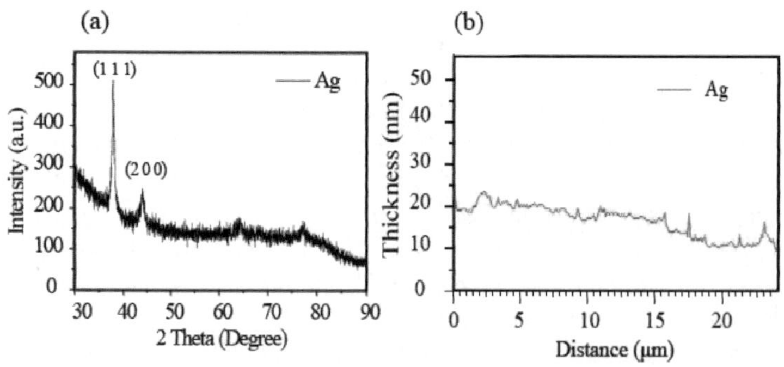

Figure 2. Characterization of stretchable silver film. (a) XRD pattern. (b)AFM image.

Figure 3 presents the scanning electron microscopy (SEM) image of stretchable nanocrack silver film on PDMS membrane. There are lots of nanocracks on the surface of the film. The initial cracks in the films distributed with random orientation uniformly (Figure 3(a)) and their length are from several to hundred nanometers. While the film is being stretched, some of the initial cracks propagate and extend into large scale to ease local stress. The typical structures of the cracks on the film being stretching are marked with red arrow in Figure 3(b). When the film is stretched thereafter, some of the initial cracks grow to larger scale cracks permanently. Figure 3(c) shows that the micro-scale cracks close when the film release, and some of the large scale cracks cannot restitution permanently. During stretching, both ends of the cracks are connected, therefore, the silver films can maintain conduction. Obviously, the crack structure is indispensable to the stretchability of the silver films.

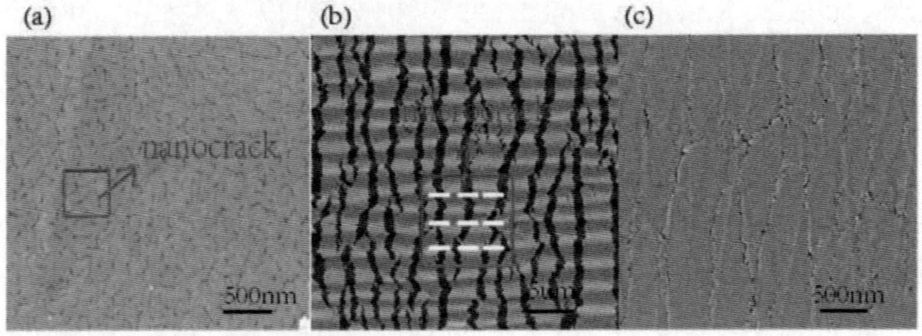

Figure 3.SEM images of a stretchable silver film on PDMS membrane (a) before; (b) under, and (c) after stretching. The red rectangles in both (a) and (b) show the typical structures of cracks.

3.2 Electrical properties of stretchable silver film

In order to tracking the maximum strain of stretchable silver film, the resistance between the two testing rectangle is measured before, under and after stretching. The stretchable silver film sample can be stretched up to the maximum strain as high as about 150% while the silver film maintains conduction (see the black curve in Figure 4(a)). Furthermore, during hundreds of stretching-releasing cycles, the resistance has been examined. Figure 4(b) shows that the value of the resistance during cyclic deformation is stable and repeatable. The maximum strain of silver films increase to 150% compared to that of the stretchable gold film in our previous work (see the red curve in Figure 4(a)), which is 120%. In addition, the resistance of the gold film is about 5 times higher than that of the silver, which indicates that the silver films have better electrical conductivity.

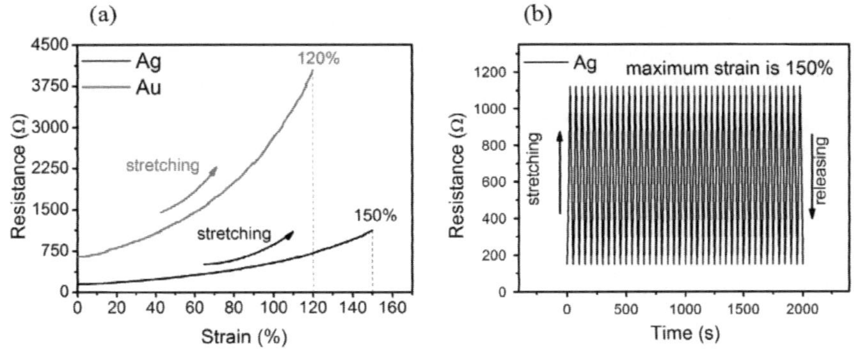

Figure4. Electrical resistance of stretchable film on PDMS membrane: (a) the compared stretchable silver film with gold film; (b) Electrical resistance of a stretchable silver film on PDMS membrane under cyclic strain of 150%. The up and down arrows lines represent the process of stretching and releasing.

4. Conclusion

In this work, we successfully fabricated the stretchable nanocrack silver films using magnetron sputtering. Results illustrate that the films can be stretched by a maximum 150% strain while maintaining great electrical conductivity. SEM images show that the initial cracks in the film distributed with random orientation uniformly. While the film is being stretched, some of the initial cracks extend into large scale. During stretching, both ends of the cracks are connected, therefore, the silver films can maintain conduction. Results reveal that, compared to gold, silver films have relatively high and better stretchability. In addition, the price of gold is about 75 times higher than that of silver, which indicates that the application of silver film is extremely extended.

Acknowledge
This work was supported by National Natural Science Foundation of China (61374013), National Program on Key Basic Research Project (973 Program, 2015CB755500), Shenzhen "Layout of Disciplines" Program on Basic Research Project (JCYJ20160331174854880), and China Postdoctoral Science Foundation 59th (162438).

References
[1] Chen, K.-L. and C.-M. Lin, *Intelligent watch*. 2015, Google Patents.
[2] Van Laerhoven, K., et al. *Medical healthcare monitoring with wearable and implantable sensors*. in *Proc. of the 3rd International Workshop on Ubiquitous Computing for Healthcare Applications*. 2004.
[3] Venkatraman, S. and S.G.J. Yuen, *Wearable heart rate monitor*. 2015, Google Patents.
[4] Larmagnac, A., et al., *Stretchable electronics based on Ag-PDMS composites.* Sci Rep, 2014. **4**: p. 7254.
[5] Cotrut, C., et al., *Mechanical, In Vitro Corrosion Resistance and Biological Compatibility of Cast and Annealed Ti25Nb10Zr Alloy.* Metals, 2017. **7**(3): p. 86.
[6] Hammock, M.L., et al., *25th Anniversary Article: The Evolution of Electronic Skin (E - Skin): A Brief History, Design Considerations, and Recent Progress.* Advanced Materials, 2013. **25**(42): p. 5997-6038.
[7] Wang, X., et al., *Silk - molded flexible, ultrasensitive, and highly stable electronic skin for monitoring human physiological signals.* Advanced materials, 2014. **26**(9): p. 1336-1342.
[8] Murphy, E.F., *The swing phase of walking with above-knee prostheses.* Journal of Rehabilitation Research & Development, 2013. **50**(9): p. xvi.
[9] Lee, J.I., et al. *Microfabricated 3D flexible tactile sensor with table-shaped structure for intelligent robot fingers*. in *Micro-NanoMechatronics and Human Science (MHS), 2010 International Symposium on*. 2010. IEEE.
[10] Chow, A.Y., *Artificial retina device*. 1991, Google Patents.
[11] Cheung, K.C., et al., *Flexible polyimide microelectrode array for in vivo recordings and current source density analysis.* Biosensors and Bioelectronics, 2007. **22**(8): p. 1783-1790.
[12] Lacour, S.P., et al., *Stretchable gold conductors on elastomeric substrates.* Applied physics letters, 2003. **82**(15): p. 2404-2406.
[13] Li, T., et al., *Stretchability of thin metal films on elastomer substrates.* Applied Physics Letters, 2004. **85**(16): p. 3435-3437.
[14] Zhou, B., et al., *Facile preparation of superhydrophobic PDMS with patternable and controllable water adhesion characteristics.* Journal of Materials Science, 2017. **52**(19): p. 11428-11441.
[15] Eyer, K., et al., *A simple low-cost method to enhance luminescence and fluorescence signals in PDMS-based microfluidic devices.* RSC Adv., 2015. **5**(17): p. 12511-12516.
[16] A.S. AiswaryaRaj. V.Biju, *Microstructure and electrical conductivity of laminated Cu/CNT/Cu composites prepared by electrode position.* Materials Science in Semiconductor Processing, **68**(2017):p. 38-47.
[17] X.G. Cao, H.Y. Zhang, Investigation into conductivity of silver-coated cenosphere composites prepared by a modified electroless process. Appl. Surf. Sci. **264**(2013): p.756-760.
[18] X.G. Cao, H Ren, H.Y. Zhang, Preparation and microwave shielding property of silver-coated carbonyl iron powder .J. Alloys Compd. **631**(2015):p.133-137.

IOP Conf. Series: Materials Science and Engineering **265** (2017) 012009 doi:10.1088/1757-899X/265/1/012009

Synthesis of Stretchable Gold Films with Nanocracks: Stretched up to 120% Strain while Maintaining Conductivity

Mei Yu, Chong Wang, Cancan Yang*, Zhe Yu

Biomedical Microdevices Research Laboratory, Shenzhen Institutes of Advanced Technology, Chinese Academy ofSciences, 1068 Xueyuan Avenue, Shenzhen 518055, China

*Corresponding Author E-mail: cc.yang@siat.ac.cn

Abstract. With the great deformability of stretch, compression, bend and twisting, while preserving electrical property, metal films on elastomeric substrates have many applications for serving as bioelectrical interfaces. However, at present, most polymer-supported thin metal films reported rupture at small elongations (<10%). In this work, highly stretchable thin gold films were fabricated on PDMS substrates by a novel micro-processing technology. The as deposited films can be stretched by a maximum 120% strain while maintaining their electrical conductivity. Electrical characteristics of the gold films under single-cycle and multi-cycle stretch deformations are investigated in this work. SEM images imply that the gold films are under the structure of nanocracks. The mechanisms of the stretchability of the gold films can be explained by the nanocraks, which uniformly distribute with random orientation in the films.

1. Introduction

Deformable electronics are being developed for many applications, including artificial blood vessel[1], electronic skins[2], electronic textiles[3], paperlike displays[4], and flexible solar cells[5]. Curved structures of conventional rigid materials have been researched and developed intensively during the past decade due to their electrical properties and simple processing technology[6]. However, freestanding metal thin films rupture when stretched beyond 1%-2% [7]. Moreover, most of biological tissues are soft, elastic and curved. Due to the large mismatch in mechanical properties, the implanted electronic devices may irritate and damage the surrounding tissue and cause immune reaction, scarring and hemorrhage[8].Currently, stretchable electronic deviceshave received extensive attentiondue to their great deformability and better Young's modulus matching with biological tissues. Metal films on elastomeric substrates can be stretched repeatedly by tens of percent over hundreds of cycles without fatigue and remain electrically conducting[9-11]. The devices are usually made from gold, semiconductors, graphene and nano-wires on stretchable substrates. However,most polymer-supported thin metal films rupture at small elongations (ε <10%), although elongations as high as 20% have been reported in few cases[12-13].

In this paper, we fabricatedhighly stretchable thin gold films with nanocrackson PDMS using magnetron sputtering. The films can maintain conducting while being stretched at large strains as about 120%. The electrical characteristics of the highly stretchable gold film are investigated,in bothsingle-cycle and multi-cycle deformation. In addition, the surface morphologies of the highly stretchable gold films on PDMS are also investigated.

2. Experiments and Results

2.1 Fabrication process of highly stretchable thin gold films on elastomeric substrates

Fabrication process of highly stretchable thin gold films on elastomeric substrates is described with the schematic illustration showed in Figure 1. A silicon waferused as a rigid backing is coated with a monolayer of 1H, 1H, 2H, 2H perfluoro-octyl-trichlorosilane (48931-10G, Sigma Aldrich) to facilitate the removal of silicone membranes at the end of fabrication. The silicone prepolymer and the cross linker are mixed in a 10:1 ratio by weight (PDMS; Sylgard 184, Dow Corning), mixing and defoaming in a conditioning mixer (AR-100, Thinky Mixer, Japan), then spun on the silicon wafer at 600 rpm, 60 s for substrate. The PDMS is cured at 80℃ for 3 h. This preparation produces a 110-μm-thick PDMS membrane for substrate. After plasma treatment on the PDMS substrate, a 3-nm-thick titanium adhesion layer followed by a 50-nm-thick gold film is deposited through a pre-patterned, stainless steel mask by a direct-current (dc) magnetron sputtering system (JS4S-75G, JSWN, China). After peeled off from the silicon wafer, the electrode sample is accomplished. For the sample, the mask pattern is designed similar to a dumbbell, two pads at the ends are squares (1.5 mm×1.5 mm), the wire in between is0.5mm in width and 8 mm in length.

Traditional electrodes are made usually by lithography process. The photoresist is spin on the substrate, in the step of exposure and development, due to the light show characteristics, part of the photoresist dissolves, the other part continues to remain in the substrate serving as the mask. After metal deposition, the photoresist residue is dissolved by stripping liquid. In this paper, the traditional process isimproved, we choose a pre-patterned stainless steel instead of photoresist serving as the mask. The stainless steel mask can be used repeatedly. The improved process saves time, cost, and improves the efficiency.

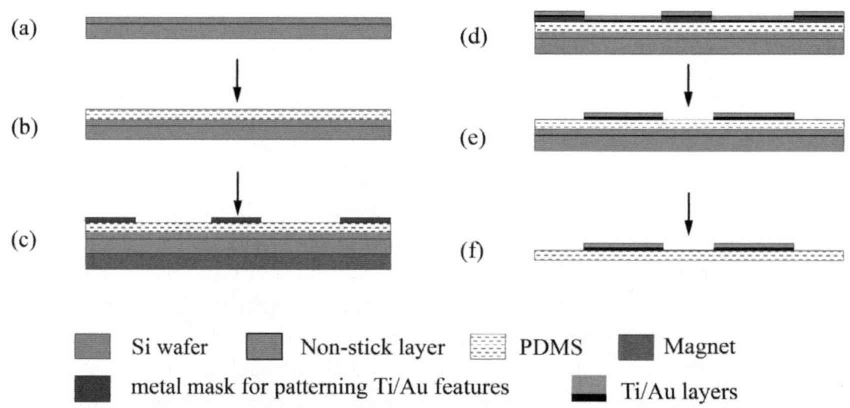

Figure 1. Fabrication process of highly stretchable thin gold films on elastomeric substrates

2.2 Electrical propertiesof highly stretchable thin gold films on elastomeric substrates

A typical electrode sample designed for this study is showed in Figure 2(a).The electrode is designed similar to a dumbbell, two pads at the ends are squares. Andthe electrical resistancesare tested between pad #1 and pad #2. A customized stretcher device designed for stretching the electrode at a certain amount of strain is also showed in Figure 2(a). Figure 2(a) shows the states of the electrode before and under stretching. From Figure 2(b), we can see that the critical uniaxial strain or the maximum uniaxial strain while the gold film maintains conduction, characterized as stretchability of the gold film, can reach as high as about 120%. The strain where the resistance is O.R. isdefined as electrical rupture strain, at which the sample is still mechanically robust. Multiple samples are tested, and the critical strain value of 120% is highlyrepeatable. Although the resistance increases along with stretching, the changes in resistance are reversible, and the resistance versus the applied strain shows a hysteresis curve. From Figure 2(b), we can also conclude that the resistance after being fully relaxed is slightly higher than that before stretching. The electrical characteristics of the highly stretchable thin gold film during multi-cycle deformation are also investigated. From Figure 2(c), we can see that the resistance of the electrode at a certain strain becomes lower and lower from one cycle to the subsequent cycles with the maximum applied strain of 110%. A continuous, cyclic deformation is applied for testing, and the shift in resistance tends to converge to a stable phase in the end.

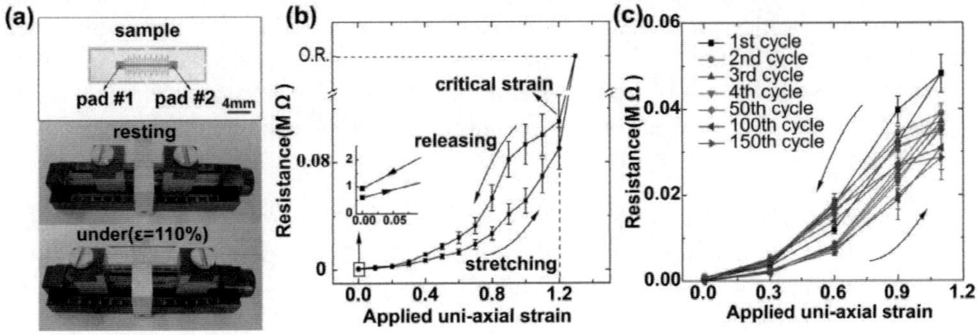

Figure 2.(a) Optical images of the 50 nm thick gold films on PDMS. The red double arrow line shows the direction of the stretch. (b) Electrical resistance of the 50 nm thick gold films on PDMS while they are stretched at different applied uniaxial strains (n=5), where O.R. represents that the resistance is out of measuring range, indicating a fracture of the film. (c) Electrical resistance of gold conductors (n=5) under cyclic applied strain when the maximum applied strain is 110%. The curved arrows present the stretching and releasing process.

2.3 The principle of stretchability of highly stretchable thin gold films on elastomeric substrates

Scanning electron microscope (SEM, NOVA NanoSEM 450, FEI) images of highly stretchable thin gold film on PDMS are showed in Figure 3. Figure 3(a) shows the SEM image of the film at the resting state. Initial nanocracks can be found in the film uniformly distributed with random orientation, and the mean length of nanocracks is 90 ± 8 nm (n =100). While the film is being stretched for the first time, some of the initial nanocracks propagate and coalesce into large microscale cracks to ease local stress, which can be seen in Figure 3(b). It is believed that due to so many initial nanocracks, gold films can have such a high stretchability. Figure 3(c) shows the SEM image of the film after stretching. We can see that the microscale cracks close when the films relax, and some turnup flakes appear in the films. Which can be concluded that, the resistance of the electrode after being fully relaxed is slightly higher than that before stretching because of the contact resistance between the two edges of cracks when the cracks close.

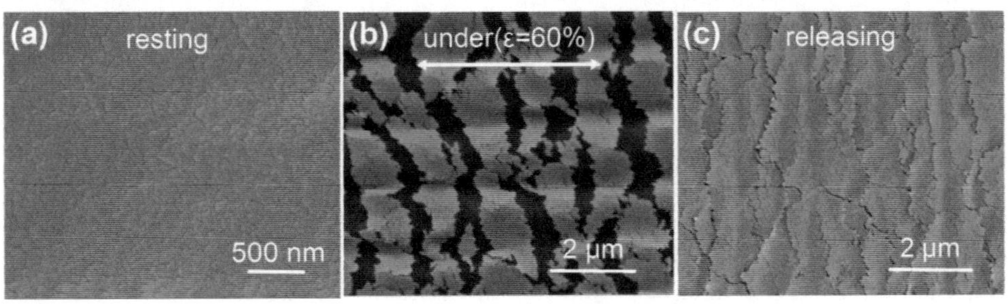

Figure 3.SEM images of the 50 nm thick gold films on PDMS before, under, and after stretching. The white double arrow line in (b) shows the direction for the uniaxial stretch.

3. Conclusion

In this work, an improved processing technology was adopted to fabricatehighly stretchable nanocrack gold films based on magnetron sputtering and a stainless steel mask, which improved the processing efficiency greatly. The gold films can be stretched as much as 120% uniaxial strain and maintain their electrical conductivity under stretching. The highly stretchable films can be bent, stretched or twisted while keeping conductive, which create a promising field in biomedical applications for its better modulus match with biological tissues and robust mechanical and electrical properties. They allow for construction of electronic integrated circuits spread over on complex and dynamic curved surfaces, creating a view for the development of stretchable electronic devices.Based on the highly stretchable gold films, we will develop stretchable microelectrode arrays,which will be designed and fabricatedusing for physiological monitoring: monitoring intracranial electroencephalography (EEG); stimulating peripheral nerves to drive muscles; monitoringepicardial electrocardiography (ECG).

Acknowledgments
This work was supported by National Natural Science Foundation of China (61374013), National Program on Key Basic Research Project (973 Program, 2015CB755500), Shenzhen "Layout of Disciplines" Program on Basic Research Project (JCYJ20160331174854880), and China Postdoctoral Science Foundation 59th (162438).

References

[1] By Peter Domachuk, K. T., Fiorenzo G. Omenetto, and David & Kaplan, L. Bio-microfluidics: Biomaterials and Biomimetic Designs. *Advanced l Materials***22**, 249-260 (2010).
[2] Lu, N., Kim, Dae-Hyeong. Flexible and Stretchable Electronics Paving the Way for Soft Robotics.*SOFT ROBOTICS***1**, 53-62, doi:10.1089/soro.2013.0005 (2014).
[3] Chen Chih-Ping, ChiangChun-Ying, YuYang-Yen, et al.High-performance, robust, stretchable organic photovoltaics using commercially available tape as a deformable substrate. *Solar Energy Materials and Solar Cells*165, 111-118 (2017).
[4] Forrest, S. The path to ubiquitous and low-cost organic electronic appliances on plastic.*Nature***428**, 911-918, doi:10.1038/nature02498 (APR 29 2004).
[5] Ghorbani Mina,Golobostanfard Mohammad Reza, AbdizadehHossein.Flexible freestanding sandwich type ZnO/rGO/ZnO electrode for wearable supercapacitor.*Applied Surface Science* 419, 277-285 (2017).
[6] Hirano L A, Escote M T, Martins-Filho L S, et al. Development of artificial muscles based on electroactiveionomericpolymermetal composites. *Artificial Organs*, 2011, 35(5): 478-483.
[7] Lacour, S. p. P., Chan, D., Wagner, S., Li, T. &Suo, Z. Mechanisms of reversible stretchability of thin metal films on elastomeric substrates. *Applied Physics Letters***88**, 204103, doi:10.1063/1.2201874 (2006).
[8] Subbaroyan J, Martin D C, Kipke D R. A finite-element model of the mechanical effects of implantable microelectrodes in the cerebral cortex. *Journal of Neural Engineering*, 2005, 2(4): 103-113.
[9] Lacour, S. P., Jones, J., Wagner, S., Li, T., Suo, Z. Stretchable interconnects for elastic electronic surfaces. *Proc. IEEE*, 2005, 93 (8), 1459–1467.
[10] ThiyagarajanKaliannan, JeongUnyong. Strategies for stretchable polymer semiconductor layers.*Mrs Bulletin*, 2017 42(2): 98-102.
[11] Lee, Y., Bae, S.,et al.Wafer-scale Synthesis and Transfer of Graphene Films. *Nano Lett.* 2010,10 (2), 490–493.
[12] Lacour, S. p. P., Wagner, S. Stretchable Gold Conductors on Elastomeric Substrates. *Applied Physics Letters* 2003 82(15): 2404-2406.
[13] P. Gruber, J. Bohm, A. Wanner, L. Sauter, et al. Nanoscale materials and modeling-relations among processing, microstructure and mechanical properties. *MRS Symposia Proceedings*, 2003, 821.

L929 cell cytotoxicity associated with experimental and commercial dental flosses

Tua-ngam P[1], Supanitayanon L[2], Dechkunakorn S[3], Anuwongnukroh N[3*], Srikhirin T[4] and Roongrujimek P[5]

[1]Research Office, Faculty of Dentistry, Mahidol University, Bangkok, Thailand
[2]Residency training program in Orthodontics, Faculty of Dentistry, Mahidol University,
Bangkok, Thailand.
[3]Department of Orthodontics, Faculty of Dentistry, Mahidol University, Bangkok, Thailand
[4]Department of Physics, Faculty of Science, Mahidol University, Bangkok, Thailand
[5]Faculty of Dentistry, Mahidol University, Bangkok, Thailand
*Corresponding author: Niwat Anuwongnukroh, E-mail: niwat.anu@mahidol.ac.th

Abstract. This aim of the study was to investigate the cytotoxicity of two commercial and two experimental dental flosses. Two commercial, Oral B® Essential Floss (nylon-waxed) and Thai Silk Floss (silk-waxed), and two experimental, Floss X (nylon-waxed) and Floss Xu (nylon- unwaxed) dental flosses were used. The cytotoxic assay was performed by using cell cultures (L929) which were subjected to cell viability test with methyl-tetrazolium. Each floss specimen (0.4 g) was placed in 1 ml of Minimum Essential Medium at 37°C with 5% CO_2 at 100% humidity in an incubator for 24 hours. After incubation, the cell mitochondrial activity was evaluated for detecting viable cells using optical density as per the guidelines of ISO 10993-5:2009(E). Cytotoxic effects were evaluated by measuring percentage of cell viability at 3 points of time- 5 mins, 30 mins, and 1 hr. The results showed that two commercial dental flosses and Floss X had cell viability about 90% at the three time points; however, the experimental Floss Xu presented 80% cell viability at 5 min and <70% cell viability at 30 min and 1 hr. The results concluded that the commercial dental flosses and the experimental dental floss with wax tested in this study were acceptable for clinical use.

1. Introduction

Conventional brushing has been found to be inadequate for the complete removal of entrapped food particles and plaque from the tooth structures. Their removal is important as it reduces the incidence of caries, gingivitis, and halitosis. The use of adjunctive cleaning aids, such as dental floss, is suggested for good dental hygiene practice as it provides easy access to interdental and proximal areas, where conventional tooth brushes may not reach. Dental flosses were traditionally made from threads which were coated with wax that provided a certain degree of lubrication, and it made the dental floss easier to be glided between the interdental areas. The wax also made the dental floss less traumatic for the patients.

Currently, dental flosses are fabricated using various polymers, such as nylon, silk, etc. [1] and vary extensively in terms of structure, size, and additives. Very few studies are conducted on the efficacy of dental floss. Although Terézhalmye et al found no clinically relevant differences in plaque-removing efficacy between woven, shred-resistant, and unwaxed dental flosses when combined with manual tooth brushing, the results were significantly better as compared to only brushing [2]. The findings were also in conjunction with a meta-review by Sälzer, where a small but significant effect on gingivitis was noted when tooth brushing was combined with flossing [3]. Although some studies have noted no additional benefit of flossing [4, 5], recommendation for its usage are often made by dental practioners. Furthermore, information should be available regarding the safety of the materials and the additives used for the fabrication of these dental flosses. Therefore, the objective of this study was to evaluate cytotoxicity associated with two experimentally fabricated dental flosses and to compare them with commercial ones.

2. Materials and Methods

Four different types of dental flosses were used in this study and were as follows:
(1). Oral B® Essential Floss, a waxed dental floss made up of nylon (Fig.1).

Fig. 1: Oral B® Essential Floss

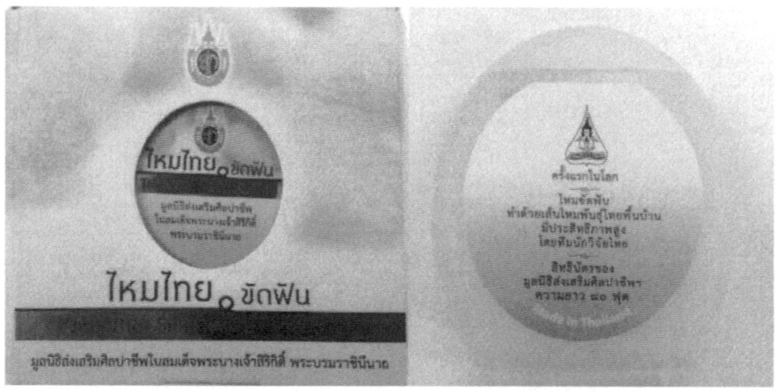

Fig.2: Thai Silk Dental Floss®

(2). Thai Silk Dental Floss®, a waxed dental floss fabricated from natural silk. It is a product of "The Foundation of the Promotion of Supplementary Occupations and Related Techniques of Her Majesty Queen Sirikit of Thailand"(Fig.2).
(3). Experimental dental Floss X, an unwaxed dental floss fabricated from nylon. It composition was evaluated with FTIR, a process that measures the range of wavelengths in the infrared region that are absorbed by a material, before subjected to further studies (Fig.3).
(4). Experimental dental Floss Xu, a waxed dental floss which was compositionally identical to Floss X but was further coated with microcrystalline wax (Fig.4).

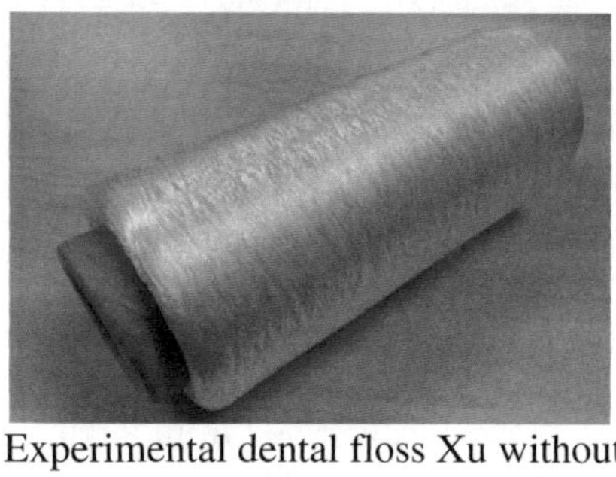

Fig.3: Experimental dental floss Xu without wax

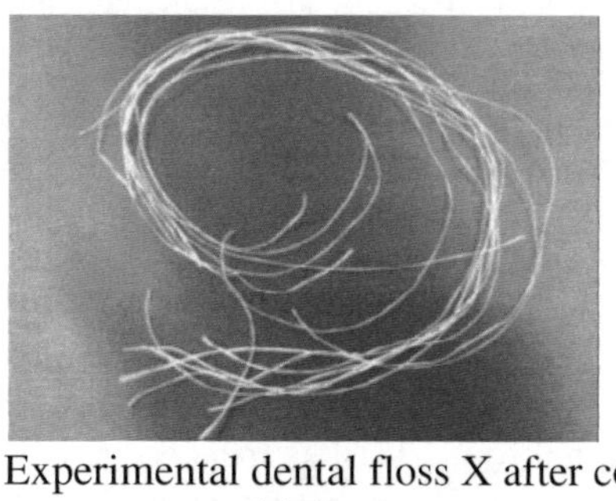

Fig.4: Experimental dental floss X after coating with wax

3. MTT assay for cell viability

The cytotoxic assay was performed using L929 cell cultures which were subjected to the cell viability test with methyl-tetrazolium. Each floss specimen (0.4g) was placed in 1 ml of Minimum Essential Medium at 37oC with 5% CO2 at 100% humidity 24 hrs in an incubator.
L929 cells were seeded into 96-well plates and maintained in culture for 24 hrs (1 doubling period) into form a semi-confluent monolayer [6]. They were then exposed to the test compound over a range of concentrations. After 24 hrs exposure, the formazan formation was determined for each treatment

concentration and compared to that determined in control cultures. For each treatment, the percentage inhibition of growth was calculated.

The cell viability was assessed at 5 mins, 10 mins, and 1 hour by tetrazolium bromide reduction (MTT) assay for mitochondrial activities in all of the test and control groups. The dental flosses were removed from the wells. The cells were cleansed three times with PBS solution. One hundred microliters of the MTT (Hi Media) solution (0.1 mg/ml of tetrazolium bromide salt dissolved in Basal Medium Eagle, GibcoBRL) was added to each well, and the plates were incubated overnight at 37 ℃ in a 5% carbon dioxide incubator.

During incubation, the yellowish extracellular MTT salt was converted into purplish intracellular formazan by metabolic enzymes in the mitochondria. Propanol in 0.04 mol/L HCL was used to lyse the Vero cells, and the purplish lysate was read using an ELISA reader (Lab Systems, Multiscan EX) with a 570-nm filter. This procedure was done for each of the four dental flosses. In addition, a similar assay was also done for control group.

Different models of dental floss were obtained and tested with elution test method onto mouse fibroblast cultures and incubated for 24 hours. Cellular viability was assessed using Dimethylthiazol diphenyltetrazolium bromide test (MTT assay) to determine the cytotoxicity level, according to MTT cytotoxicity test: ISO 10993-5: 2009(E)[7]. Cellular viability was assessed at 3 periods of time: 5 minutes, 30 minutes, and 1 hour. The lower the viability % value, the higher the cytotoxic potential of the test item. If viability was reduced to <70% of the blank, it was considered to have potential cytotoxicity.

4. Results

The results of the cytotoxicity test for the 4 dental flosses were shown in Table 1.

Table 1: The percentage of cell viability of dental floss at each period of time

Types of Dental Floss	Time	Cell Viability [%]
Oral B	5 min	93.41
	30 min	91.32
	1 hour	85.45
Floss X	5 min	93.68
	30 min	90.76
	1 hour	88.26
Floss Xu	5 min	87.26
	30 min	62.44
	1 hour	56.33
Thai Silk	5 min	103.67
	30 min	93.71
	1 hour	90.04

For all tested dental flosses, the highest percentage of cell viability was observed at 5 min and the least was at 1 hr. The percentage of cell viability decreased when the time was increased. The experimental dental floss Xu had the least and Thai silk had the highest percentage of cell viability. The floss Xu at 30 mins and 1 hr showed cell viability of <70%.

5. Discussion

Biocompatibility of materials is an important consideration for patients, clinicians, laboratory technicians, and manufacturers, especially if the material is intended to used intraorally. Therefore, it is of

upmost importance that the material is devoid of any toxic, leachable, or diffusible substances that can be absorbed into the circulatory system, as they can cause local and systemic responses, including teratogenic or carcinogenic effects. As a result, cytotoxic studies should be conducted before dental materials are selected for use [8]. By definition, the cytotoxicity of an agent means the toxicological risks caused by a material or its extract in a cell culture [9].

A variety of commercial dental flosses are available in waxed, unwaxed, monofilament, or multifilament forms. Monofilamentous dental floss coated with wax can easily glide between teeth and does not fray, but it is generally higher in cost as compared to uncoated ones. However, studies have shown that there is no difference in terms of effectiveness between waxed and unwaxed dental floss, but some have added antibacterial agents such as chlorhexidine [10]. Factors to consider in choosing a floss include the space between teeth and user preference. Dental tape is also a type of floss that is wider and flatter than conventional floss and recommended for larger tooth surface area [8] [11].
The experiments in this study were performed according to the ISO 1999, which leaves some flexibility in specimen fabrication. Two aspects require further standardization for a more precise and reliable evaluation of the biological effects of dental materials. These are the ratio of the specimen surface/cell culture medium and the specimen fabrication which have been partly addressed in ISO (2008). L-929 fibroblasts and gingival fibroblasts have previous been shown to have similar cytotoxicity levels. L-929 fibroblasts make a useful screening model for in vitro toxicity testing of dental materials [12] because of its excellent reproducibility. Therefore, the L-929 cell line was preferred to primary gingival fibroblasts in this study.
The cytotoxicity results of the two commercial and two experimental dental flosses used in this study showed that the experimental unwaxed dental floss (Floss X) had the least of percentage of cell viability, but the other waxed dental flosses had similar percentage of cell viability. This could have been due to two possible factors. Firstly, the two commercial dental flosses (Oral B and Thai silk) were sealed in marketing packages before distribution for sales. Therefore, it can be assumed that they were sterilized, and the experimental dental floss with wax (Floss Xu) required coating by hot melting wax, which could have bactericidal effects on any adherent micro-organisms. On the contrary, the unwaxed experimental dental floss (Floss X) did not undergo any sterilization procedure and was only sterilized prior to the cytotoxicity test. Furthermore, the amount of specimen by weight for cytotoxicity test was more for Floss X than the other dental flosses. These factors could have contributed to higher toxicity of Floss X as compared to the other dental flosses.
Studies have shown that the ability of different types of dental floss to remove dental plaque does not vary significantly [13]; the least expensive floss has essentially the same impact on oral hygiene as the most expensive one. Factors to be considered when choosing the right floss or whether the use of floss is appropriate as interdental cleaning aid may be based on [11]:

- Tightness of the contact area; to determine the width of floss
- Contour of gingival tissues
- Roughness of the interproximal surface
- Manual dexterity and preference of the patient: to determine if supplemental device is required.

Although flossing is commonly used to disrupt the oral biofilm between the teeth and to prevent gingival diseases, its effectiveness is greatly determined by patient's preference, technique, and motivation to floss daily. Flossing is considered to be a more difficult method of interdental cleaning than using an interdental brush. Interdental brushes are preferred due to their one-handed usage and time efficiency as compared to flossing [14]. A groove in the gingival margin, known as a floss cleft, can also form after repeatedly using floss incorrectly along the mesial and distal surfaces of the tooth [15]. Traumatic flossing immediately after the placement of amalgam fillings can also result in an amalgam tattoo [16]. Moreover, the both waxed and unwaxed dental flosses are also found to have common problems due to misuse of product. But the benefits of flossing in lowering the risk of gingivitis and possibly interproximal dental caries outweigh their drawbacks.

6. Conclusion

Flossing is essential aid for oral health maintenance as the toothbrush bristles may not reach all the surfaces in between teeth or under gum line. However, materials used for the fabrication of dental flosses should be non-toxic and safe for usage. The results of the cytotoxicity tested conducted on L929 fibroblast cell line using MTT assay showed that the two commercial dental flosses and Floss X had cell viability of approximately 90% at 5 mins, 30 mins and 1 hour following exposure to the material. However, the experimental Floss Xu presented a lower cell viability, observed at <70% at 30 min and 1 hr. As per the results of the study the two commercial dental flosses and the waxed experimental dental floss were considered to be acceptable for clinical use.

References

[1] ISO Standards catalogue of oral care product, ISO/TC 106/SC 7, for Integrated dental floss and handles ISO28158:2010.
[2] Terézhalmy GT, Vartizek RB, Biesbrock AR. Plaque-Removal Efficacy of Four Types of Dental Floss. J Periodontol 2008;79:245-251.
[3] Sälzer S, Slot DE, Van der Weijden FA, Dörfer CE. Efficacy of inter-dental mechanical plaque control in managing gingivitis – a meta-review. J Clin Periodontol 2015; 42 (Suppl. 16): S92–S105.
[4] Berchier CE, Slot DE, Haps S, Van der Weijden GA. The efficacy of dental floss in addition to a toothbrush on plaque and parameters of gingival inflammation: a systematic review. Int J Dent Hyg 2008;6:265-279.
[5] Sambunjak D, Nickerson JW, Poklepovic T, Johnson TM, Imai P, Tugwell P, Worthington HV. Flossing for the management of periodontal diseases and dental caries in adults. Cochrane Database Syst Rev. 2011;7:CD008829.
[6] Coecke S, Balls M, Bowe G, Davis J, Cstraunthaler G, Hartung T, Hay R, Merten O, Price A, Schectman L, Stacey G and Stokes W. Guidance on Good Cell Culture Practice, A report of the second ECVAM task force on good cell culture practice. ATLA 2005;33:261-287
[7] ISO 10993-5:2009(E), Biological evaluation of medical devices -- Part 1: Evaluation and testing within a risk management process,
[8] Hanks CT, Wataha JC, Parsell RR, Strawn SE. Delineation of cytotoxic concentration of two dentin bonding agents in vitro. J Endod 1992;18:589-596.
[9] Hanks CT, Strawn SE, Wataha JC, Parsell RR, Craig RG. Cytotoxic effects of resin components on cultured mammalian fibroblast. J Dent Res 1991;70:1450-1455.
[10] Muniz FWMG, da Silva Lima H, Rösing CK, Martins RS, Moreira MMSM, Carvalho RS. Efficacy of an unwaxed dental floss impregnated with 2% chlorhexidine on control of supragingival biofilm: A randomized, clinical trial. J Invest Clin Dent 2017;e12280. doi: 10.1111/jicd.12280
[11] Darby M, Walsh M. Dental Hygiene Theory and practice.(2010). Saunders Elsevier (3rd Ed).pp 402
[12] Schedle A, Samorapoompichit P, Rausch-Fan XH, Franz A, Füreder W, Sperr WR, Sperr W, Ellinger A, Slavicek R, Boltz-Nitulescu G, Valent P. Response of L-929 fibroblasts, human gingival fibroblasts and human tissue mast cells to various metal cations. J Dent Res 1995;74:1513-1520
[13] Heasman P (editor). (2008). Restorative dentistry, paediatric dentistry and orthodontics (2nd Ed.). Edinburgh: Churchill Livingstone.pp.37.
[14] Imal P, Yu X, MacDonald D. Comparison of interdental brush to dental floss for reduction of clinical parameters of periodontal disease: A systematic review. Can J Dent Hygiene 2012;46:63-78.
[15] Hallmon W, Waldrop T, Houston G, Hawkins B. Flossing Clefts: Clinical and Histological Observations. J Periodontol 1986;57:501-504.
[16] Neville BW, Damm DD, Allen CA, Bouquot JE. (2002). Oral & maxillofacial pathology (2nd Ed.). Philadelphia: W.B.Saunders.pp.269-272.

IOP Conf. Series: Materials Science and Engineering **265** (2017) 012011 doi:10.1088/1757-899X/265/1/012011

Preparation of a novel breviscapine-loaded halloysite nanotubes complex for controlled release of breviscapine

Min Gao[1*], Liqian Lu[1], Xiaoyue Wang[1], Houke Lin[1], Qingsong Zhou[1, 2]

[1] School of Pharmacy, Jiangxi Science and Technology Normal University, Nanchang 330013, China

zhoushuimo@163.com

Abstract. For sustain the release rate and prolong half-life of breviscapine in vivo, the breviscapine-loaded halloysite nanotubes complex was prepared. The breviscapine was encapsulated into halloysite nanotubes (HNTs) using a vacuum process. The complex were investigated by scanning electron microscopy (SEM), differential scanning calorimetry (DSC), transmission electron microscope (TEM), X-ray diffraction (XRD) and fourier transform infrared spectroscopy(FT-IR). The formation of breviscapine-loaded HNTs complex was proved by the test results of SEM, DSC, TEM and IR analysise. The results confirmed that breviscapine was successfully loaded in the halloysite nanotubes. Additionally, the in vitro drug release of breviscapine from breviscapine-loaded HNTs complex was investigated, the result indicated this complex has apparent sustained-release effect.

1. Introduction

Breviscapine (4, 5, 6- tetrahydroxyflavone-7-O-glucuronide, Figure 1), a cerebrovascular drugs extracted from the Chinese herb *Erigeron breviscapinus (Vant.) Hand.-Mazz*, has been frequently used to clinically treat cerebrovascular diseases such as cerebral thrombosis, cerebral infarction, and cerebral circulation insufficiency [1]. However, One of the limitations of breviscapine is its short half-life in vivo, therefore, it was necessary to develop a sustained-release dosage form with good bioavailability and stable efficacy to reduce the frequency of breviscapine administration [2].

Halloysite nanotubes (HNTs) is a natural aluminosilicate nanotube from a clay mineral with a similar structure to kaolinite, which has a predominantly hollow nanotube structure with an inner diameter of 5-10nm [3-4]. HNTs have attracted attentions for their potential as a new type of nanotube container for encapsulating various active agents such as drugs, marine biocides, antifouling paint, cosmetics, and other functional agents [5]. HNTs is a viable and inexpensive nanotube container with a hollow tubular structure, which has high aspect ratio, specific surface area, excellent chemical and thermal stability.

HNTs' special lumen can be used for loading small molecules and monomer [6-7]. Many researches of HNTs as drug carrriers have been reported, indicating the nanocomposites have controlled release properties under certain conditions. Based on this thinking, halloysite as a drug carrrier to load brevescapine, a sustained release solid dispersions loading has been investigated in this work.

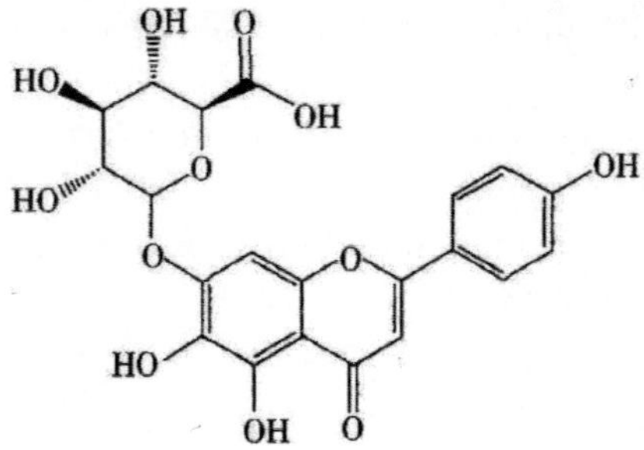

Figure 1. The chemical constituent of breviscapine.

2. Materials and methods

2.1 Materials

The following materials were obtained from their respective suppliers: brevescapine (Shanxi Kingsci Biotechnology Co. Ltd); Reference substances of brevescapine (Shanghai Jin sui Biological Technology Co. Ltd, the purity is 98%); halloysite (SIGMA-ALDRICH); Methanol (HPLC pure; Xilong Chemical Co. Ltd); acetonitrile (HPLC pure; Xilong Chemical Co., Ltd); Potassium hydrogen phosphate anhydrous (AR, Tianjin Damao Chemical Reagent Factory); Disodium phosphate dodecahydrate (AR, Tianjin Daimao Chemical Reagent Factory); Sodium dihydrogen phosphate (AR, Tianjin Daimao Chemical Reagent Factory).

2.2 Preparation of brevescapine-loaded HNT complex

Brevescapine was loaded into HNTs according to the procedure described by Price et. al. [8], with slight modifications. Brevescapine (6mg) was added to 10 mL of absolute methanol and vortexed for 5min. Then, 60mg of HNTs was thoroughly dispersed into the solution with a magnetic stirrer for 20min to form a suspension. The suspension was kept in a vacuum chamber at 25°C. After 20 min, the vacuum was stopped and air was allowed into the chamber for 10 min. This process was repeated four times to increase the encapsulation efficiency. The air in the lumen of HNTs was replaced with brevescapine solution through the vacuum process. Next, brevescapine-loaded HNTs complex were separated from the solution by centrifugation using a high speed centrifuge (Neofuge15R) at 7000 rpm for 15 min. The tubules were washed by using deionic water to remove unloaded brevescapine. Finally, brevescapine-loaded HNTs complex was dried at 45°C in an oven. The dried complex was pulverized in a mortar and left to the next experiment [9].

2.3 Testing and characterization

2.3.1 Scanning Electron Microscopy. The surface morphology of the pristine HNTs and the brevescapine-loaded HNTs complex was examined respectively through instrument Field Emission Scanning Electron Microscopy (FE-SEM, Zeiss, sigma200, Germany) at an accelerating voltage of 20 kV. Before SEM observation, the sample was deposited on a brass hold and sputtered with gold [10].

2.3.2 Transmission Electron Microscopy. For transmission electron microscopy (Tecnai G2 20 S-TWIN) observation, pristine HNTs and the brevescapine-loaded HNTs complex were immersed in ethylalcohol and dispersed with ultrasonic waves for 40min. A dilute suspension of sample was dropped on the copper grid and left undisturbed for 1 min. Then, excess solution was removed using filter paper. Finally, the sample was detected after the grids was dried.

2.3.3 Fourier transform infrared spectroscopy. Fourier transform infrared spectroscopy can find out the chemical stability of drug in the formulation. Pristine HNTs, pristine breviscapine, physical mixture of HNTs and breviscapine, and the brevescapine-loaded HNTs complex also analyzed respectively using the Fourier-transform infrared (FT-IR, Bruker Vertex 70) at 25°C. Each sample was prepared using the potassium bromide (KBr) pellets method. Briefly, each sample was ground and mixed with KBr, and prepared as a pellet by compression under a force of 10tons in a hydraulic press. All FT-IR spectra's were obtained wavelengths 400 to 4000 cm^{-1} with a resolution of 1cm^{-1}.

2.3.4 Differential scanning calorimetry. Differential scanning calorimetry (DSC) were performed on different samples to determine their composition and predict the thermal stability. DSC measurements of pristine HNTs, pristine breviscapine, physical mixture of HNTs and breviscapine, and the brevescapine- loaded HNTs complex were done using a DSC Q20 (made in USA). The heating-cooling- heating run for all samples were recorded in the temperature range from 0 to 400°C at a scan rate of 10°C/min under nitrogen atmosphere.

2.3.5 X-ray diffraction . XRD of pristine HNTs, pristine breviscapine, physical mixture of HNTs and breviscapine, and the brevescapine-loaded HNTs complex were performed over the XRD-6100 at the incident X-rays (λ=1.54nm) and scanning range of 10–80°. It equipped with a continuous scan attachment and a proportional counter, with Ni-filtered Cu Kα radiation.

2.4 Drug content

Breviscapine loading was determined by dispersing accurately weighed amounts of sample (3mg) in 10mL of absolute methyl alcohol. Then, the samples were broken using an ultrasonic homogenizer for 30 min. After centrifugation was carried out at 7000 rpm for 10 min to separate broken HNTs out of suspension, the supernatant was filtrated through a 0.45μm membrane, and the amount of breviscapine was determined in triplicate for each sample using HPLC. Quantification of breviscapine was carried out using a calibration curve [11].

2.5 In vitro drug release studies

In vitro dissolution of complex was carried out using the dissolution apparatus equipped with six paddles. The dissolution rates were measured at 37°C under 50 rpm paddle speed. Weighed quantities of each sample were placed in buffer solutions. A 100 mL of phosphate buffer solution (pH6.8) was used to simulate the intestinal environment. A 1 mL of aliquot was taken from the vessel at definite intervals of time (5min, 10min, 15min, 30min, 45min, 1, 2, 3, 4, 5, 6 and 8h) and replaced with an equal volume of corresponding dissolution medium. These samples were analyzed using HPLC and concentration of breviscapine was calculated using the calibration curves constructed from the reference standards. The in vitro release studies were performed in triplicate for each of the samples.

3. Results and discussion

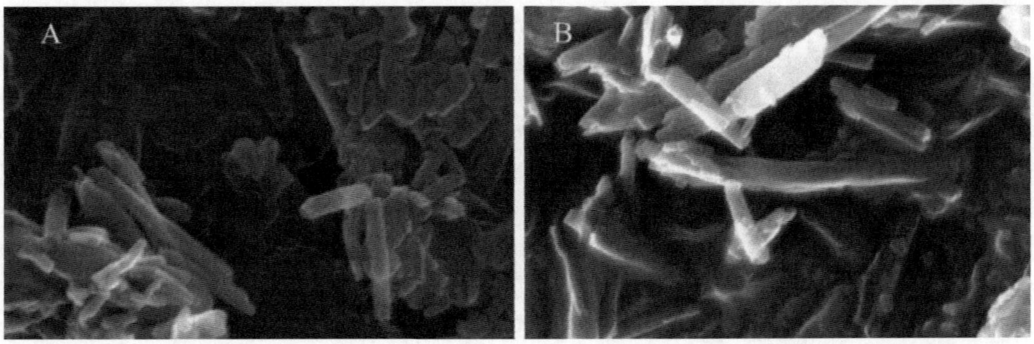

Figure 2. SEM images of (A) pristine HNTs, (B) brevescapine-loaded HNTs complex.

The surface morphology of each sample was investigated by SEM. The image of pristine HNTs in Figure 2(A) showed a rod shape and indicated that HNTs have a length of approximately 1 μm. This result corresponded to those of previous studies. Figure 2(B) showed brevescapine-loaded HNTs complex. The brevescapine-loaded HNTs complex exhibited a smooth surface similar to the pristine HNTs, which proved that breviscapine has no adsorbed on the surface of complex.

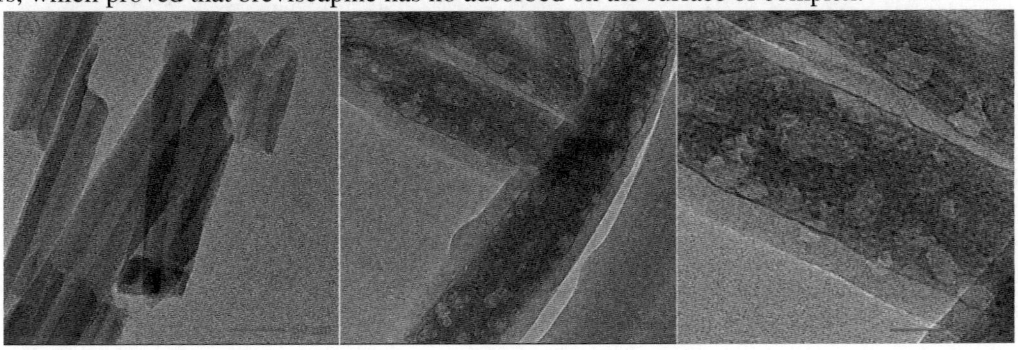

Figure 3. TEM images of (A) pristine HNTs, (B) brevescapine-loaded HNTs complex.

TEM was used to further characterize the structural changes of the pristine HNTs and brevescapine-loaded HNTs complex. Figure 3(A) showed the image of the Breviscapine/HNT particle, indicating that the HNTs have typical hollow structures with an outer diameter of 50–60 nm and a lumen diameter of 10–20 nm. Figure 3(B) showed that brevescapine-loaded HNTs complex. Testifying breviscapine was successfully loaded in halloysite, which was consistent with the SEM images.

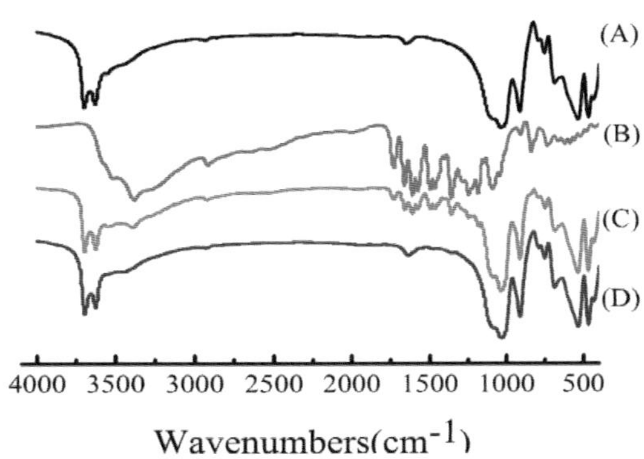

Figure 4. FT-IR spectra of (A) pristine HNTs, (B) pristine breviscapine, (C) physical mixture of HNTs and breviscapine, and (D) brevescapine-loaded HNTs complex

To confirm the structure of the HNTs and the existence of brevescapine-loaded HNTs complex, FT-IR studies were conducted (Figure 4). In the FT-IR spectrum of the pristine HNTs (Figure 4A), the peaks appeared at 3696 and 3621 cm^{-1} were assigned to the OH stretching vibration of the inner Al-OH groups of the HNTs. The peaks at 1645, 911, and 470 cm^{-1} are assigned to the stretching vibrations of OH groups in absorbed water, the OH deformation vibration of the inner hydroxyl groups, and Si-O-Si deformation vibration, respectively. Figure 4(B) displayed the FT-IR spectrum of the breviscapine. The characteristic band at 3379cm^{-1} was observed and was due to OH stretching. A sharp peak observed at 2916 and 1148cm^{-1} was due to CH stretching. The FT-IR spectra of breviscapine showed a peak at 1727cm^{-1}, which was ascribed to C=O stretching. The peak observed at 1660cm^{-1} was C=C aromatic stretching. The peak at 1247 cm^{-1} characterizes the stretching vibration of the epoxy. Figure 4(C) depicted the FT-IR spectrum of the physical mixture of breviscapine and halloysite, it can be seen from the figure that the spectrum of the mixture is the overlap of the infrared spectra of halloysite and breviscapine. The peaks of breviscapine from 4000cm^{-1} to 400cm^{-1} can be observed in the spectrum. No additional peaks were seen, indicating there was no interaction of breviscapine either with halloysite. Figure 4(D) showed the FT-IR spectrum of brevescapine-loaded HNTs complex after encapsulated breviscapine. As can see in the spectrum, the peak of brevescapine-loaded HNTs complex is almost similar to the HNTs peak. The characteristic peak of breviscapine has no test in the complex, indicating that the surface of HNTs almost no residual breviscapine, breviscapine were almost loaded to the HNTs cavity [12].

In order to further verify the formation of the drug loaded, the DSC thermograms of pristineHNTs (A), pristinebreviscapine (B), physical mixture of HNTs and breviscapine (C), brevescapine-loaded HNTs complex (D) were obtained (Figure. 5).

Differential scanning calorimetry (DSC) was used to study the crystalline behaviour of HNTs and brevescapine-loaded HNTs complex. Figure 5(A) shows the crystallization exotherms of breviscapine. The drug of breviscapine showed a sharp peak at 186.59°C and 354.14°C due to its exothermic peak and melting peaks. Figure 5(B) shows the crystallization

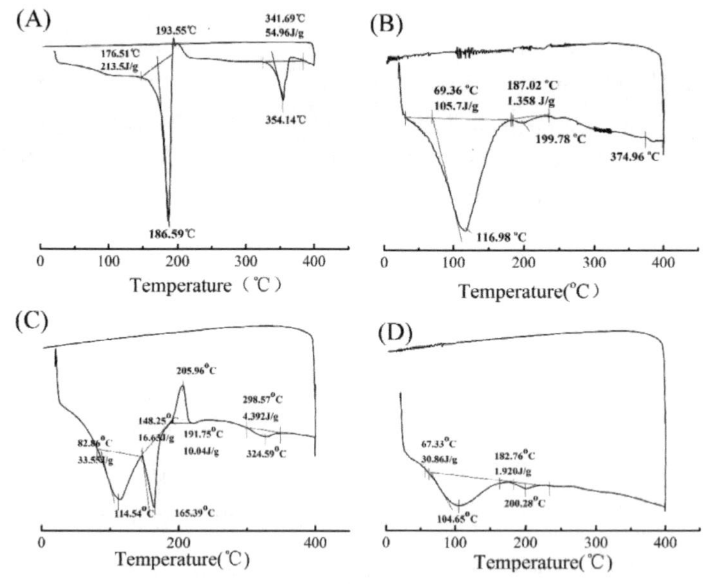

Figure 5. DSC thermograms of (A) pristine breviscapine,(B) pristine HNTs, (C) physical mixture of HNTs and breviscapine, and (D) brevescapine-loaded HNTs complex.

exotherms of HNTs. The apparent endothermic peaks at 116.98°C and 374.96°C are the dehydration peak and the melting peak of halloysite. Figure 5(C) shows the crystallization exotherms of HNTs and breviscapine mixture. The peaks of breviscapine can be observed in the DSC thermogram. No extra peaks were seen, indicating that there was no interaction of breviscapine either with halloysite. Figure 5(D) shows the crystallization exotherms of brevescapine-loaded HNTs complex. The drug of breviscapine showed a sharp peak at 186.59 °C and 354.14°C due to its exothermic peak and melting peaks, which did not appear in the DSC thermogram of brevescapine-loaded HNTs complex. The DSC results prove that the breviscapine were almost in the cavity of HNTs [13].

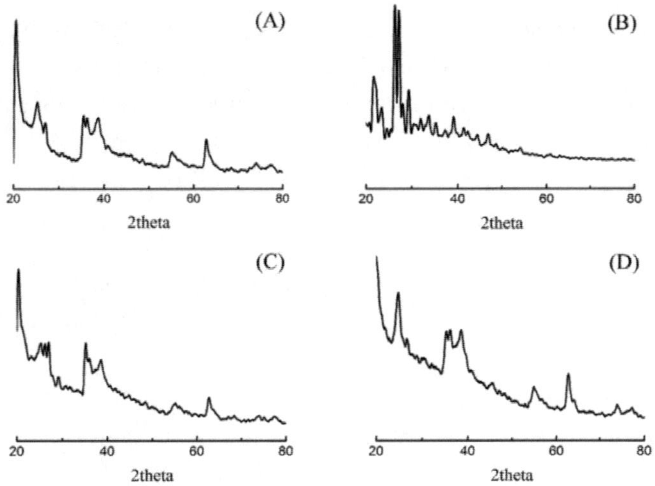

Figure6. XRD spectra of (A) pristine HNTs,(B) pristinebreviscapine , (C) physical mixture of HNTs and breviscapine, and(D) brevescapine-loaded HNTs complex.

X-ray diffraction method commonly used in the identification of crystalline compounds, each having a different crystalline material interplanar spacings at the same angle, thus showing different diffraction peaks. XRD spectrum of HNTs is shown in Figure 6(A). The main absorption peak of halloysite nanotubes is about 20.06, 29.92 at 2θ. XRD spectrum of breviscapine is shown in Figure 6(B). The main absorption peak of brevescapine is about 27, 28 at 2θ. XRD spectrum of HNTs and breviscapine mixture is shown in Figure 6(C), The diffraction angle of halloysite and breviscapine are both shown in the figure 6(C), that can see that halloysite and breviscapine is physical mixture. Figure 6(D) shows the XRD spectrum of brevescapine-loaded HNTs complex, its diffraction angle is similar to halloysite, it has no breviscapine's diffraction angle. It can be determined that the load was formed [14].

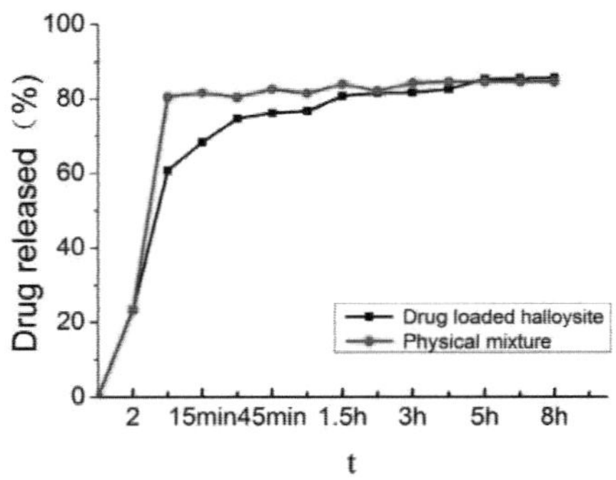

Figure 7. Cumulative release of breviscapine from (a) physical mixture of HNTs and breviscapine, (b) brevescapine-loaded HNTs complex at 37°C.

Figure 7(a) is a graph of the release profile of breviscapine from the physical mixture of HNTs and breviscapine. In this sample, the breviscapine is observed to release in a fairly rapid manner over a period of 3min, and the breviscapine has been completely released rapidly. Figure 7 (b) is a graph of the release profile of breviscapine from the HNTs complex. Breviscapine from the halloysite nanotubes in the 30min release of 80%. Compared with the physical mixture, the drug loaded halloysite has a apparent sustained-release effect [15].

4. Conclusion

In summary, a novel brevescapine-loaded HNTs complex was successfully prepared by using a vacuum process. The morphology and structure were characterized by SEM and FT-IR. Further characterizations were verified by TEM, DSC, XRD. The in vitro release experiment confirmed the brevescapine-loaded HNTs complex can develop a sustained-release and stable efficacy. In addition, HNTs are environmental friendly and low cost materials that are very easy to handle. Therefore, HNT containing breviscapine have great potential for drug in the nanoscopic drug carriers.

Acknowledgments

This work was supported by National Natural Science Foundation of China (No. 81160397)

Reference

[1] Jiang L, Xia Q, Dong X, et al. Neuroprotective effect of breviscapine on traumatic brain injury in rats associated with the inhibition of GSK3β signaling pathway[J]. *Brain Research*, 2017, 1660: 1-9.

[2] Guo C, Zhu Y, Weng Y, et al. Therapeutic time window and underlying therapeutic mechanism of breviscapine injection against cerebral ischemia/reperfusion injury in rats[J]. Journal of ethnopharmacology, 2014, 151(1): 660-666.

[3] Lvov Y M, Shchukin D G, Mohwald H, et al. Halloysite clay nanotubes for controlled release of protective agents[J]. AcsNano, 2008, 2(5): 814-820.

[4] Rooj S, Das A, Thakur V, et al. Preparation and properties of natural nanocomposites based on natural rubber and naturally occurring halloysite nanotubes[J]. Materials & Design, 2010, 31(4): 2151-2156.

[5] Abdullayev E, Price R, Shchukin D, et al. Halloysite tubes as nanocontainers for anticorrosion coating with benzotriazole[J]. ACS applied materials & interfaces, 2009, 1(7): 1437-1443.

[6] Marney D C O, Russell L J, Wu D Y, et al. The suitability of halloysite nanotubes as a fire retardant for nylon 6[J]. Polymer degradation and stability, 2008, 93(10): 1971-1978.

[7] Handge U A, Hedicke-Höchstötter K, Altstädt V. Composites of polyamide 6 and silicate nanotubes of the mineral halloysite: influence of molecular weight on thermal, mechanical and rheological properties[J]. Polymer, 2010, 51(12): 2690-2699.

[8] RR Price, BP Gaber, Y Lvov. In-vitro release characteristics of tetracycline HCl, khellin and nicotinamide adenine dineculeotide from halloysite; a cylindrical mineral[J]. Journal of microencapsulation, 2001, 18(6):713-722.

[9] Yendluri R, Otto D P, De Villiers M M, et al. Application of halloysite clay nanotubes as a pharmaceutical excipient[J]. International Journal of Pharmaceutics, 2017, 521(1):267-273.

[10] Liu H Y, Du L, Zhao Y T, et al. In vitro hemocompatibility and cytotoxicity evaluation of halloysite nanotubes for biomedical application[J]. Journal of Nanomaterials, 2015, 16(1): 384.

[11] Ding L, Stilwell J, Zhang T, et al. Molecular characterization of the cytotoxic mechanism of multiwall carbon nanotubes and nano-onions on human skin fibroblast[J]. Nano Letters, 2005, 5(12): 2448-2464.

[12] Kulkarni V H, Kulkarni P V, Keshavayya J. Glutaraldehyde - crosslinked chitosan beads for controlled release of diclofenac sodium[J]. Journal of Applied Polymer Science, 2007, 103(1): 211-217.

[13] Kubade P, Tambe P. Influence of halloysite nanotubes on morphology, crystallization, mechanical and thermal behaviour of PP/ABS blends and its composites in presence and absence of dual compatibilizer[J]. Composite Interfaces, 2016, 23(5): 433-451.

[14] Gorrasi G. Dispersion of halloysite loaded with natural antimicrobials into pectins: Characterization and controlled release analysis[J]. Carbohydrate polymers, 2015, 127: 47-53.

[15] Lee M H, Park H J. Preparation of halloysite nanotubes coated with Eudragit for a controlled release of thyme essential oil[J]. Journal of Applied Polymer Science, 2015, 132(46).

IOP Conf. Series: Materials Science and Engineering **265** (2017) 012012 doi:10.1088/1757-899X/265/1/012012

Initial Tensile and Residual Forces of Pigmented Elastomeric Ligatures from Various Brands

Wassana Wichai[1, a], Niwat Anuwongnukroh[2, b] *, Surachai Dechkunakorn[2, c], Rattiporn Kaypetch[1], d and Peerapong Tua-ngam[1,e]

[1]Research Office, Faculty of Dentistry, Mahidol University, Bangkok, Thailand
[2]Department of Orthodontics, Faculty of Dentistry, Mahidol University, Thailand

*Corresponding author: Niwat Anuwongnukroh, E-mail: niwat.anu@mahidol.ac.th

Abstract. This study aimed to investigate the initial tensile and residual forces of pigmented elastomeric ligatures (clear, pink, and metallic) from three commercial brands – Brand 1 (USA), Brand 2 (USA), and Brand 3(China). Twelve elastomeric ligatures of each brand and color were evaluated for initial tensile and residual forces after stretching for 28 days at 37°C by a Universal Testing Machine. The results showed that the highest initial tensile force was 14.78 N, 20.71 N, and 15.1 N for the metallic color of Brand-1, pink color of Brand -2, and metallic color of Brand -3, respectively. There were significant ($p<0.05$) differences in the initial tensile force of each brand, except clear and metallic color of Brand-1 & 3 and pink color of Brand-2 & 3. Similarly, among the pigmented ligatures from each brand, significant ($p<0.05$) differences were observed in the initial tensile force, except metallic color of Brand-1 & 3. Brand-3 had the highest residual force after 28 days, whereas the loss of force was 80-90% in Brand-1 & 2 and 20-30% in Brand-3. There were also significant ($p<0.05$) differences in the residual forces in each color and brand, except metallic color of Brand-1. In conclusion, there were significant differences in the initial tensile and residual forces among the three pigmented elastomeric ligatures of the three commercial brands.

1. Introduction

Since their first introduction in orthodontics more than three decades ago, elastomeric modules have gained universal acceptance by the profession. They are made up of polyurethane, which is a generic term associated with elastic polymers with a urethane linkage [1] and are used in the form of ligatures and chains or modules [2] to ligate archwires to orthodontic bracket. They have key benefits such as low cost, easy application, reduced chair time, patient comfort, and satisfaction.

Securing archwires with a single elastomeric module produces forces of 50-150 g [3]. The force exerted by an elastomeric ligature depends on the initial force and force rate. These elastomers have two important properties- tensile strength (TS) and extension to TS. The TS can be defined as the maximum force a material can bear before breakage and rupture. An extension to TS is measured as the extension at which maximal stress occurs [4].

Over the years, there has been an increased patient demand for esthetically pleasing appearance during treatment with fixed appliances, promoting the need for esthetic accessories. To meet these demands, many companies have produced elastomeric ligatures in different colors. There have been some studies that have reported differences in mechanical characteristics between clear and pigmented elastomeric chains, but few studies have focused on elastomeric ligatures [5,6]. Therefore, the purpose of this in vitro study was to investigate elastomeric ligatures from three commercial brands available in Thai market with 3 different colors (clear, pink, and metallic) for their initial tensile force and force loss after stretching for 28 days.

2. Materials and Methods

Clear, pink, and metallic pigmented orthodontic elastomeric ligatures from 3 commercial brands, Brand-1 (C-1, Dynaflex USA), Brand-2 (C-2, Ormco USA) and Brand-3(C-3, W&H China) were used in this study. All samples were categorized into 9 groups according to the brands and colors (Table 1). The tensile property was tested according to ISO 21606:2007, which measured the initial tensile force and, then, the residual force after stretching the ligatures for 28 days at 37°C. For each test, twelve modules of each brand and color were randomly selected.

2.1 Initial tensile force

The module was stretched using two half rods (radius 0.5 mm.) on a Universal Testing Machine (Instron 5566, Instron Corp., Buckinghamshire, UK, England) and extended at a cross-head speed of 100 mm/minute to 0.155 inches (diagonal length of upper central incisor bracket) and held for 5 seconds. The force exerted was determined in Newtons at 30±2 seconds after reaching the latter extension, following which the results of the initial extension force of each brand were compared. This test was performed at 23±2°C at a relative humidity of 50±10%.

2.2 Residual force

The module was stretched using two stainless pin at 0.155 inches distance (diagonal length of upper central incisor bracket) on a supporting plate. All samples are stored in water (ISO 3696:1987, Grade 3) at 37±2°C for 28 days. The support plate was then removed maintaining the extended condition of the samples and immediately placed in water (ISO 3696:1987, Grade 3) at 23±2°C for 30±2 mins. The samples on the support plate was removed from water and transferred to the Universal Testing Machine. The residual force was determined in Newtons at 0.155 inches (diagonal length of upper central incisor bracket) exerted at 23±2°C. Residual forces were calculated as the percentage of the initial extension force using the following formula-

$$\text{Percentage of residual force} = \frac{residual\ force\ x100}{initial\ extension\ force}$$

3. Statistical analysis

Two-way ANOVA was used to test the effect of brand and color on initial extension force. Multiple linear regression was employed to test the effect of brand and color on percentage of residual force. Multiple comparison by Bonferroni's test was used to compare the measurements among the groups. The level of statistical difference was accepted at p-value<0.05.

4.Results

From the results for the tensile force (Table 1), the metallic pigmented ligatures of C1 and C3 had the highest initial tensile force followed by pink and clear colors, respectively. However, the pink color of C2 had the highest initial tensile force followed by metallic and clear colors, respectively. The percentage of force loss after stretching for 28 days was 78-84% for C-1, 79-89% for C-2, and 19-27% for C-3. There were significant ($p<0.05$) differences in the initial tensile force among the brands in each color, except clear and metallic colors of C1-C2 and pink color of C2-C3 (Table 2). Considering the pigmented ligatures of each brand (Table 3), significant ($p<0.05$) differences were noted in the initial tensile force among the pigmented ligatures of each brand, except pink-metallic color of C1 and C3. There were also significant ($p<0.05$) differences in the residual tensile force among the brands for each color (Table 4). For the pigmented ligatures of each brand (Table 5), there were significant

($p<0.05$) differences in the residual tensile force among the pigmented ligatures of each brand, except pink and metallic color of C1.

Table 1. Means of initial extension and residual forces at 28 days

Brands	N	group	color	Initial		Residual force		%
				X	S.D	X	S.D	Force Loss
C1	12	1	Clear	11.05	0.76	2.11	0.13	80.9
		2	Pink	11.22	0.43	2.38	0.25	78.79
		3	Metallic	14.78	3.47	2.32	0.19	84.3
C2	12	4	Clear	10.83	0.55	2.27	0.03	79.04
		5	Pink	15.10	0.85	1.54	0.04	89.8
		6	Metallic	13.63	0.29	2.14	0.05	84.3
C3	12	7	Clear	8.38	0.82	6.74	0.59	19.57
		8	Pink	19.61	1.50	14.26	1.16	27.28
		9	Metallic	20.71	1.71	15.96	1.31	22.94

Table 2. Differences in initial tensile force based on brands and colors

Initial Tensile Force									
Color	Clear			Pink			Metallic		
Brand	C1	C2	C3	C1	C2	C3	C1	C2	C3
C1	-	NS	S	-	S	S	-	NS	S
C2	NS	-	S	S	-	NS	NS	-	S
C3	S	S	-	S	NS	-	S	S	-

Table 3. Differences in initial tensile force based on brands and colors

Initial Tensile Force									
Brand	C1			C2			C3		
Color	Clear	Pink	Metallic	Clear	Pink	Metallic	Clear	Pink	Metallic
Clear	-	S	S	-	S	S	-	S	S
Pink	S	-	NS	S	-	S	S	-	NS
Metallic	S	NS	-	S	S	-	S	NS	-

Table 4. Differences in residual tensile force on based on brands and colors

Residual Tensile Force									
Color	Clear			Pink			Metallic		
Brand	C1	C2	C3	C1	C2	C3	C1	C2	C3
C1	-	S	S	-	S	S	-	S	S
C2	S	-	S	S	-	S	S	-	S
C3	S	S	-	S	S	-	S	S	-

Table 5. Differences in residual tensile force based on brands and colors

Residual Tensile Force									
Brand	C1			C2			C3		
Color	Clear	Pink	Metallic	Clear	Pink	Metallic	Clear	Pink	Metallic
Clear	-	S	S	-	S	S	-	S	S
Pink	S	-	NS	S	-	S	S	-	S
Metallic	S	NS	-	S	S	-	S	S	-

IOP Conf. Series: Materials Science and Engineering **265** (2017) 012012 doi:10.1088/1757-899X/265/1/012012

5. Discussion

This study compared 3 brands of polyurethane-based elastomeric modules which were made in two different countries with a wide variety of mechanical properties [7]. Overall, the ligature quality and composition is a result of a combination of technology, technique refinement, and quality of the raw materials used during manufacturing [1,8] but this information are generally withheld by the manufacturers [7,9]. When exposed to the oral environment for a lengthy period of time, these ligatures absorb water and saliva, which disrupts the internal connections in the polymer network, thereby, promoting permanent deformation [9].

The testing methods employed in this study followed ISO21606:2007 that was designed to simulate the clinical situation of extended elastomeric ligature by using a calibrated Universal Testing Machine. Although there have been some studies on tensile strength (TS) of elastomeric ligature at breaking point [4,10,11], it may not be applied in clinical practice because the elastomeric ligatures are only extended around the diameter of the bracket to engage archwire with bracket wings. Indirect measurement of the structure and mechanical properties of these polymers provides insight into the loss of force. The results of this study showed that all elastomeric ligatures decreased in tensile force after 28 days. Percentage of force loss in 28 days were 78-84% for C1, 79-89% for C2, and 19-27% for C3. These results were in confirmation with previous studies that presented similar pattern of force decay and 50-60% of force loss during the first 24-hour interval [1,12,13]. C1 and C2 were nearly consistent with these reported values and can be considered as clinically acceptable when compared to commercially available products. However, C3 had a higher tensile force even after 28 days.

The results also found that there were significant differences among each color of each brand. The addition of pigments to elastomeric ligature by manufactures to increase patient acceptance might have affected their mechanical properties, esp. in terms of initial force and residual force after 28 days. In a previous study, Lam et al. found that the addition of coloring materials to ligatures effected tensile strength. The filler materials added for tinting of the elastomeric chains can affect their polymer network and, hence, their properties, as shown by Anthony and Paulose [14], who noted less force delivery and greater force decay in pigmented elastomeric chains as compared to non-pigmented ones and that these values differed according to the color and the manufacturer. Moreover, Renick and colleagues [6] observed significant differences in the glass transition temperatures of gray and purple pigmented elastomeric chains. Taloumis et al [1] found a negative correlation between the inner diameter and force. In addition, Boonyarangkul [15] noted a positive correlation between cross-section thickness and initial extension force. However, this paper did not measure the size of elastomeric ligatures.

The findings from this study provided data of a simulated clinical condition, that may be different from the forces provided by elastomeric ligatures in the oral cavity. In vivo observations of force delivery from modules may reveal lower amount of force due to pH and temperature variations in the oral environment, along with the accumulation of plaque and formation of microbial colonies on elastomeric modules. Ash and Nikolai [13] found that force degradation is more rapid in water and oral environment than in air. In addition, the decay in the oral environment become statistically greater than in water after 10 days.

Assessments of the physical properties of elastomeric ligatures have reported significant changes in the structure and composition of these materials after exposure to the oral environment [1,16]. After exposure to the oral environment for 3 weeks and subsequent analysis by optical microscopy and spectroscopy, the elastomeric ligatures exhibited precipitates of calcium phosphate on the surface. Significant changes were found in the structure and composition of their surface following exposure to the oral environment, indicative of the severity of changes that can occur in the properties of these materials [16]. In this study, there were differences in the tensile forces among the brands and colors which may have resulted due to variations among the manufacturers in the processing technique, composition, and size of elastomeric ligatures.

6. Conclusion

Significant differences were observed in the initial tensile and residual forces among the three pigmented elastomeric ligatures tested in this study from three different manufacturers.

References

[1] Taloumis LJ, Smith TM, Hondrum SO, and Lorton L. Force decay and deformation of orthodontic elastomeric ligatures. Am J Orthod Dentofacial Orthop. 1997;111:1-11.
[2] Eliades T, Eliades G, Watts DC, Watts DC, Brantley WA. Elastomeric ligatures and chains. In: Brantley WA, Eliades T. eds. Orthodontic Materials: Scientific and Clinical Aspects. Stuttgart: Thieme. 2001:174-187.
[3] Sim AP, Waters NE, Birnie DJ, Pethybridge RJ. Acomparison of the forces required to produce tooth movement in vitro using two self-ligating brackets and pre-adjusted bracket employing two types of ligation. Eur J Orthod. 1993;15:377-85.
[4] Lam TV, Freer TJ, Brockhurst PJ, and Podlich HM. Strength decay of orthodontic elastomeric ligatures. J Orthod. 2002;29:37-43.
[5] Lu TC, Wang WN, Tarng TH, Chen JW. Force decay of elastomeric chain a serial study. Part II. Am J Orthod Dentofacial Orthop. 1993 Oct;104(4):373-7.
[6] Renick MR, Brantley WA, Beck FM, Vig KWL, Webb CS. Studies of orthodontic elastomeric modules. Part 1: Glass transition temperatures for representative pigmented products in the as-received condition and after orthodontic use. Am J of Orthod Dentofacial Orthop. 2004;126(3):337-43.
[7] Wong AK. Orthodontic elastic materials. Angle Orthod. 1976;46(2):196-204.
[8] Souza EV, Mendes AM, Almeida MAO, Quintao CCA. Percentual de degradacao das forcas liberadas por ligaduras elasticas. Rev Dental Press Ortod Ortop Facial. 2008;13(2):138-45.
[9] Martins MM, Mendes AM, Almaida MAO, Goldner MTA, Ramos VF, Guimaraes SS. Estudo comparative entre as diferentes cores de ligaduras elasticas. Rev Dental Press Ortod Ortop Facial 2006;11(4):81-90.
[10] Evangelista MB, Berzins DW, Monaghan P. Effect of disinfecting solutions on the mechanical properties of orthodontic elastomeric ligatures. Angle Orthod. 2007;77(4):681-7.
[11] Dowling PA, Jones WB, Lagerstrom L, Sandham JA. An investigation into the behavioural characteristics of orthodontic elastomeric modules. Br J Orthod. 1998;25(3):197-202.
[12] Sriamporn T, Thamronganankul N, Na Ranong N, Swasdison S. Physical and mechanical properties of commercial orthodontic elastomeric ligatures. Chula Dental Journal. 2009;32:11-22.
[13] Ash JL, Nikolai RJ. Relaxation of orthodontic elastomeric chains and modules in vitro and in vivo. J Dent Res. 1978;57(5-6):685-90.
[14] Antony PJ, Paulose J. An in-vitro study to compare the force degradation of pigmented and non-pigmented elastomeric chains. Indian J Dent Res. 2014;25(2):208-13.
[15] Boonyarangkul T, Anuwongnukroh N, Dechkunakorn S, Suddhasthira T, Laokijcharoen P. Thai orthodontic elastomeric ligatures. Mahidol Dental Journal. 2010;30(2):141-51.
[16] Eliades T, Eliades G, Watts DC. Structural conformation of in vitro and in vivo aged orthodontic elastomeric modules. Eur J Orthod. 1999;21(6):649-58.

Effect of the Addition of 3% Co in NiTi Alloy on Loading/Unloading Force

Phukaoluan A[1], Dechkunakorn S[2*], Anuwongnukroh N[2], Khantachawana a[1,] Kaewtathip P[1], Kajornchaiyakul j[3], and Wichai W[4]

[1]Department of Mechanical Engineering, King Mongkut's University of Technology Thonburi,Tungkru, Bangkok 10140, Thailand
[2]Department of Orthodontics, Faculty of Dentistry, Mahidol University, Rajthevee, Bangkok 10400 Thailand
[3]National Metal and Materials Technology Center, 114 Thailand Science Park, Pathumthani 12120 Thailand
[4]Research Office, Faculty of Dentistry, Mahidol University, Rajthevee, Bangkok 10400 Thailand

*Corresponding author: Surachai Dechkunakorn, E-mail: surachai.dec@mahidol.ac.th

Abstract. The study evaluated the loading-unloading force in the load-deflection curve of the fabricated NiTiCo and NiTi wires. Wire alloys with Nickel, Titanium, and Cobalt (purity-99.95%) with atomic weight ratio 47Ni:50Ti:3Co and 50.6Ni:49.4Ti were prepared, sliced, and cold-rolled at 30% reduction, followed by heat treatment in a furnace at 400oC for 1 hour. The specimens of wire size of 0.016 x 0.022 inch2 were cut and subjected to three-point bending test to investigate the load-deflection curve at deflection point 0.25, 0.5, 0.75, 1.0, 1.25, and 1.5 mm. Descriptive statistic was used to evaluate each variables and independent t-test was used to compare between the groups. The results presented a load-deflection curve that resembled a typical superelastic wire. However, significant differences were seen in the loading-unloading forces between the two with an average loading force of 412.53g and 304.98g and unloading force of 292.40g and 208.08g for NiTiCo and NiTi wire, respectively. The force at each deflection point of NiTiCo in loading-unloading force was higher than NiTi wire. This study concluded that the addition of 3%Co in NiTi alloy can increase the loading-unloading force of NiTi wire but were within the range for orthodontic tooth movement.

1. Introduction

In fixed orthodontic treatment, archwires play a crutial role for the delivery of force required for tooth movement. They are able to delivery vertical (intrusion and extrution), horizontal (buccal and lingual), and torque forces (bucco-lingual and linguo-buccal). The efficiency of these forces is dependent on the quality, type, and size of wire. Nowadays, there are many types of wires used in orthodontic treatment and are generally imported or exported to other countries with different quality and price. NiTi (nickel-titanium) wires have found wide applications in orthodontics due to their exceptional mechanical properties but may vary based on composition, transformation temperature range, and loading and unloading forces [1-2]. The so-fabricated NiTi alloys used in this study, 50.6%Ni:49.4%Ti (NiTi), 46%Ni:49%Ti:5%Cu (NiTiCu), and 47%Ni:50%Ti:3%Co (NiTiCo) have suitable mechanical properties and superelastic behavior as commercial orthodontic wires. Studies have also considered compositional changes in the NiTi wires to control or develop potentially better wire properties. For eg. Cu has been added to decrease the hysteresis temperature [3], and Co to increase the corrosion resistance [4-5]. These wires, Ni:Ti; Ni:Ti:Cu, and Ni:Ti:Co, had an austenite finish (Af) temperature close to 37°C which gave superior superelastic behavior without residual permanent strain after unloading [6]. NiTiCo wire is reported to have adequate mechanical properties and improved corrosion resistance [7-8]; however, the current literature rarely present data on the loading and unloading force, which is importance if these wires are to be clinically used for the orthodontic tooth movement. The mechanical property of wire used in orthodontics should be optimized before clinical utilization. The possible differences in the components of alloys can change their mechanical properties and, consequently, their clinical performance. Therefore, the aim of this study was to investigate the loading and unloading force of the so-fabricated wires using load-deflection curve wih a three-point bending test for their possible application in orthodontic tooth movement.

2. Materials and Methods

Ingots of Nickel (purity 99.9%), Titanium (purity 99.8%), and Cobalt (purity 99.95%) were first cleaned in acid (HF:HNO3:H2O=1:4:5), following which 47.0% Nickel, 50.0%Titanium and 3.0%Cobalt (Ni:Ti:Co) and 50.6%Ni:49.4%Ti (Ni:Ti) atomic weight elements were melted and mixed to form alloy ingots by vacuum arc melting method in argon gas. After homogenization at 800o for 60 min, the cast ingots were sliced into thin bars with 1.5 mm thickness using an EDM wire cutting machine, cold-rolled at reduction 30%, and annealed at 400oC for 60 min. Then, the polished thin bars were cut and finely polished with sand paper for the fabrication of NiTiCo and NiTi wire specimens with dimensions - 0.016 x 0.022 inch2 (0.4064 x 0.5588 mm2).

Three-point bending test was performed by a universal testing machine (UTM-Instron) for each of the three so-fabricated NiTiCo and NiTi wire specimens to obtain a load-deflection curve for assessing the loading and unloading forces. For the bending test, a wire span wire of 10 mm between supports was used. The specimen wires were loaded to a maximum deflection of 1.5 mm with 1 KN load cell with the deflection rate 5 mm/min at 36+1oC in a water bath. The loading and unloading forces were reported at the deflection points of 0.25, 0.5 0.75, 1.0, 1.25, and 1.5 mm from the load-deflection curve (Fig 1-2). Descriptive statistics were used to evaluate each variables and independence t-test was used to analyze the differences between each deflection point.

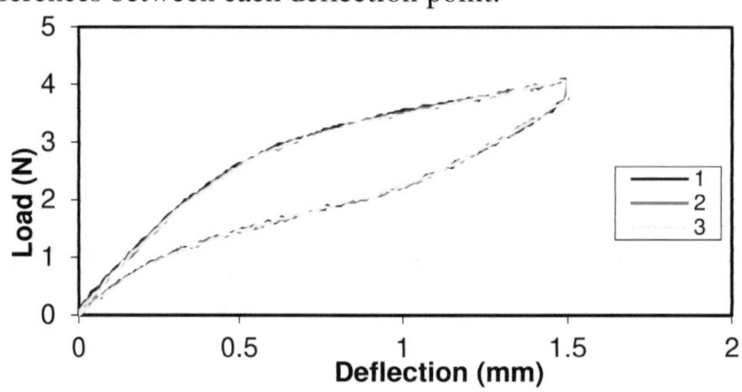

Fig 1: Load-deflection curve of the three fabricated NiTi wire specimens

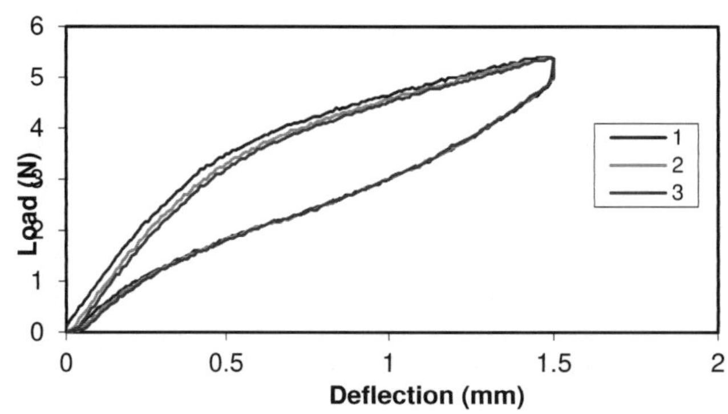

Fig 2. Load-deflection curve of the three fabricated NiTiCo wire specimens

3. Resuts

The results showed that the mean forces were 426.10 g (range 186.15- 545.70 g) for loading and 304.80 g (range 508.47- 106.59g) for unloading of NiTiCo wire and 304.98 g (range 157.08- 435.54 g) for loading and 208.08 g (range 412.08- 100.98 g) for unloading of NiTi wire (Table 1-3). There were significant ($p<0.05$) differences in the loading and unloading forces between NiTiCo and NiTi wirse. The NiTiCo wire had a higher force at each deflection point and the differences between each deflection point was more than NiTi wire (Table 2-5). Figure 1 and 2 showed the load-deflection curves of the fabricated NiTi and NiTiCo wires respectively (3 specimens each), and were notably different in the 'plateau region' of loading and unloading curves.

Table 1: Mean and standard deviation of the fabricated NiTi and NiTiCo wires in loading and unloading forces

	Loading				**Unloading**			
	NiTi		**NiTiCo**		**NiTi**		**NiTiCo**	
N	**Mean**	**SD**	**Mean**	**SD**	**Mean**	**SD**	**Mean**	**SD**
1	286.62	1.08	464.51	0.19	191.76	0.92	335.38	0.15
2	314.16	1.25	410.04	0.14	210.12	1.04	289.34	0.06
3	312.12	1.24	403.75	0.12	222.36	1.07	289.68	0.17
Total	304.98*	1.19	426.10*	0.15	208.08*	1.01	304.80*	0.13

* significant difference at $p<0.05$

Table 2: Mean and standard deviation of loading and unloading forces of the fabricated NiTi wire at each deflection point

NiTi	**Deflection**											
	0.25		**0.5**		**0.75**		**1.00**		**1.25**		**1.50**	
	Mean	**SD**	**Mean**	**SD**	**Mean**	**SD**	**Mean**	**SD**	**Mean**	**SD**	**Mean**	**SD**
Loading	157.08	0.02	278.46	0.09	347.82	0.21	388.62	0.22	415.14	0.23	435.54	0.23
Unloading	100.98	0.02	170.34	0.18	207.06	0.19	250.92	0.22	319.26	0.22	412.08	0.23

Table 3: Mean and standard deviation of loading and unloading forces of the fabricated NiTiCo archwire at each deflection point

NiTiCo	**Deflection**											
	0.25		**0.5**		**0.75**		**1.00**		**1.25**		**1.50**	
	Mean	**SD**	**Mean**	**SD**	**Mean**	**SD**	**Mean**	**SD**	**Mean**	**SD**	**Mean**	**SD**
Loading	186.15	9.38	331.5	7.21	408.00	4.33	463.59	3.61	508.47	2.16	545.70	2.88
Unloading	106.59	2.16	184.11	0.72	243.27	2.61	306.0	0.0	390.15	2.16	508.47	2.16

Table 4: Mean and standard deviation of loading and unloading forces between each deflection point of the fabricated NiTi wire

Force	**Diff. between each deflection point**				
	0.5-0.25	**0.75-0.5**	**1.0-0.75**	**1.25-1.0**	**1.5-1.25**
Loading Force	121.38	69.36	40.80	26.52	20.40
Unloading Force	69.36	36.72	43.86	68.34	92.82

Table5: Mean and standard deviation of loading and unloading forces between each deflection point of the fabricated NiTiCo wire

Force	**Diff. between each deflection point**				
	0.5-0.25	**0.75-0.5**	**1.0-0.75**	**1.25-1.0**	**1.5-1.25**
Loading Force	142.46	73.78	52.7	45.9	32.64
Unloading Force	76.84	57.46	63.24	84.32	131.24

4. Discussion

This study used three-point bending test to evaluate the loading and unloading forces of fabricated wires according to the recommendations made by the American Dental Association and Standard No.32 from the American Nation Standards Institute. Three-point bending test is an appropriate mechanical testing method for beams, like orthodontic wires, to evalulae inactivation (loading) and deactivation (unloading) forces [9]. During this test, the induced stresses may increase the transformation temperature range, as described by the Clausius-Clapayron relationship [10]. Superelastic behavior application of the stress led to a linear increase in strain up to the point where the martensitic transformation started to occur during loading. This plateau-phase was due to formation of stress-induced martensitic transformation (SIM). In unloading, a relationship exists between stress and strain down to the point where the austenitic transformation starts to occur. The unloading plateau increases until the restoration of austenitic structure is complete.

Evans and Durning classified orthodontic alloys into five categories: 1) stainless steel, gold; 2) first stabilized NiTi; 3) superelastic wires – active austenites with TTR below intraoral temperature (pseudoelastic): 4) thermodynamic wire- active martensite with TTR close to intraoral temperature; 5) a group of thermodynamic wires featuring diverse TTR (e.g. NiTiCu35oC, 37oC, 40oC) [11]. According to the TTR of the wires used in this study (the TTR Af37oC-Ms 36oC of the fabrication Ni:Ti and Af39oC-Ms 37oC of the fabrication Ni:Ti:Co) [12], the Ni:Ti wire may be classified under Category 4- active martensite and Ni:Ti:Co may be classified under Category 5. Moreover, both wires would be classified as thermodynamic wires. The scientific basis enabling the use of thermoelastic arch wires are various and accounts the argument that bone remodels itself more effectively if subjected to a dynamic load rather than a static one [13-14]. The thermoelastic properties feature is the so-called "shape memory effect", which has remarkable clinical applications. Through deflection and repeated temperature cycles, the wire in the austenitic phase is able to memorize a preformed shape, including specific orthodontic archforms. By lowering the temperature, the alloy is transformed into martensite and becomes pliable and easily deformed. However, every time the temperature rises above Af to the austenitetic phase, the wire will remember and recover the ideal archform.

The forces of orthodontic wire, which are applied to the crowns of the teeth, are distributed over the entire supporting structure and so are the stresses and strains. From a cellular point of view, distribution of stress (force per unit area), distortion of the periodontal ligament (shear stress, strain), and bone deformation (strain) are critical factors, and the remodeling response is directly related to stress and strain levels within the periodontium [15]. The orthodontic force as an extrinsic mechanical stimulus, thus, evokes a biologic cellular response that aims to restore equilibrium by remodeling of the periodontal supporting tissue [16-17]. Because it is very difficult to measure stresses and strains within the periodontal ligament of loaded teeth directly, measurement of only the forces that are applied directly to teeth with known root surface areas can provide an estimate of these parameters.

Load-deflection properties of archwires, which are considered the most important parameters determining the biological nature of tooth movement, are obtained by classic three-point bending tests. For rectangular wires, these tests are conducted such that the force is applied perpendicular to the ribbon side of archwire. The unloading curve is the curve of interest for orthodontic tooth movement. In superelastic archwires, this curve is characterized by an almost horizontal unloading plateau, which shows that these NiTi archwires are able to exert constant forces in certain range of deflection [18-22]. For this study (Fig. 1 & 2), the unloading plateaus were not parallel to the horizontal plane, which showed that these two fabricated NiTi wires did not exert the constant force in the range of deflection.

In orthodontics, an optimal approach should result in the highest possible rate of tooth movement without irreversible damage to the periodontal ligament, the alveolar bone, or the root. Nikolai [23] defined optimum force as the force that yielded the maximum desirable biologic response with minimum tissue injury producing rapid tooth movement with small or no clinical discomfort. In order to move teeth, it is necessary to apply an orthodontic force, which produce a pressure above a dental root film capillary blood pressure of about 15g/cm2 and below 20 g/cm2, onto the dental root film (US Patent 5759029). Thus, the magnitude of the optimal orthodontic force required is normally with the range of 0.5 to 3 N. This relatively low force should be applied continuously in order to achieve correction of teeth. Such force may reduce the potential for patient discomfort, tissue hyalinization, and undermining resorption. Hence, the ideal wire should behave elastically and be able to produce light and continuous forces over the period of use.

In this study presented much higher force of NiTiCo and NiTi (Table 1-5); however, the forces of orthodontic tooth movement should depend on which tooth to move, type of tooth movement, and length of the root. A previous study reported the numerical force for optimal tooth movement in different root length and found 110-130 g. of force for mandibular canine movement, 130-170g for maxillary canine movement, 270-320g for mandibular first molar movement, and 320-360g for maxillary first molar movement in the long root length group [24]. Each deflection point of this study (Table 2-3) showed that the fabricated wire had unloading at deflection point range 0 - 1.25 mm for NiTi and range 0 - 1.0 mm for NiTiCo in the range of tooth movement. Therefore, these fabricated NiTi wires could be used in adjusting teeth irregularity of not more than 1.00 mm.

For achieving optimal forces, it is essential to select an appropriate diameter and to eliminate rectangular archwires from the leveling phase. Trials evaluating superelastic and conventional NiTi archwires with regard to the movement speed of a tooth as well as the pain sensation have not validated the differences in their properties [25]. The superelastic austenitic NiTi significantly reduces the need for replacing thin archwires with the thicker for leveling and alignment [26].

Clinically, there is no single archwire that possesses all of the desired qualities necessary for all stages of orthodontic treatment. A straight superelastic leveling archwire is a highly complex, statically in determinate system, and it is impossible to predict exactly the force system acting on each tooth. The excessive forces exerted by NiTi archwire may lie far beyond biologically safe limits [27-29]. In orthodontic treatment, there are many phase of treatment and many levels of teeth irregulality. In early leveling phase of the treatment, if the malocclusion has severe teeth irregularity, the treatment should be started with the small diameter wires which have more flexibility for loading force into the bracket to move the teeth to correct the teeth in proper position by unloading force.

5. Conclusion

The study presented the load-deflection curve of two fabricated orthodontic rectuangular wires (in size diameter 0.016 x 0.022 inch2), namely NiTi and NiTiCo. Significant differences in the loading-unloading forces were noted with the addition of 3% Co in NiTi composition. Both wires had high loading and unloading force in each deflection point, thus, they should be used in later leveling phase or in teeth with mild irregularity.

References

[1] Hirokazu Nakano, et al., 1999, Mechanical properties of several nickel-titanium alloy wire in three point bending test. Am J Orthod and Dentofac. Orthop. p 390-395.

[2] Isarapatanapong R., et al., 2008, The comparison between the properties of various commercial NiTi archwires used in orthodontics.The 2nd Thailand Metallurgy Conference. p.101-106.

[3] Batalu D. etal., 2008, A review on TiNi shape memory alloys (SMA) use for medical applications. Recycling aspects, Universitatea Politetehnica din Bucuresti, Romania, p.1- 9

[4] Wang QY. and Zheng YF. 2008, The electrochemical and surface analysis of $Ti_{50}Ni_{47.2}Co_{2.8}$ alloy for orthodontic use. Dent Materials, p. 1207-1211.

[5] Hosoda H. et al.,1997, Martensite transformation temperatures and mechanical properties of ternary NiTi alloys with off stoichiometric composition. Tohoku University, Japan, p.291-301.

[6] Phukaoluan A. Khantachawana A, Dechkunakorn S, Anuwongnukroh N, Santiwong P, and Kajornchaiyakul J. Improvement of mechanical and biological properties of TiNi alloys by addition of Cu and Co to orthodontic archwires. International Orthodontics 2016;14:295-310.

[7] Phukaoluan A. Khantachawana A, Dechkunakorn S, Anuwongnukroh N, Santiwong P, and Kajornchaiyakul J. Effect of Cu and Co Additions on Corrosion Behavior of NiTi alloys for Orthodontic Applications. Advanced Materials research 2012, Vols 378-379; p650-654.

[8] Phukaoluan A. Khantachawana A, Dechkunakorn S, Anuwongnukroh N, Santiwong P, and Kajornchaiyakul J. Influence of Co Addition on Mechnical Behavior of TiNi for Orthodontic Application. The Second TSME International Conference on Mechanical Engineering 2011(Oct.), Krabi, Thailand.

[9] NSI/ADA, Council on Dental Material and Devices. Specification(1977) No.32 for Orthodontic wires.
[10] Iijima M., Ohmo H., Kawashima I., Endo K., Brantley MA and Mizoguchi I. Biomaterials. Vol23(2002).p. 1769.
[11] Evans TJW. and Durning P.: Orthodontic products update. Br. J Orthod. 1996,23,1-4, 269-275.
[12] Phukaoluan A. Khantachawana A, Dechkunakorn S, Anuwongnukroh N, Santiwong P, and Kajornchaiyakul J. Influence of Cold Work on Properties of SMAs for Madical Devices. 2nd Internation Conference on Smart Materials and Nanotecnology in Engineering Lecture Notes in Information Technology, Vol20, p 270-275,2012
[13] Dalatra M., Melsen B.: Does the transition temperature of Cu-NiTi archwires affect the amount of tooyh movement during alignment? Orthod Craniofac. Res. 2004,7(1),21-25
[14] Damon DH.: The Damon low-friction bracket: a biologically compatible straight wire system. J clin. Orthod.1998,32,670-680.
[15] Melsen B. Biological reaction of alveolar bone to orthodontic tooth movement.Angle Orthod. 1999;69:151-158.
[16] Robert WE, Goodwin WC, Heiner SR. Cellular response to orthodontic force. Dent Clin North Am. 1981;25:3-18.
[17] Yoshikawa K. Biomechanical principles of tooth movement. Dent Clin North Am.1981;25:19-26.
[18] Parvizi F, Rock WP. The load/deflection characteristics of thermally activated orthodontic archwires. Eur J Orthod 2003;25:417-21.
[19] Gatto E, Matarese G, Di Bella G, Nucera R, Borsellino C, Cordasco G. Load-deflection characteristics of superelastic and thermal nickel-titanium wire. Eur J Orthod 2013;35:115-23.
[20] Lombardo L, Marafioti M, Stefanoni F, Molloca F, Siciliani G. Load deflection characteristics and force level of nickel titanium initial arecwires. Angle Orthod 2012;82: 507-21
[21] Nuccera R, Gatto E, Borsellino C, Acclo P, Fabioano F, Matarese G, et al. Influence of bracket-slot design on the forces released by superelastic nickel-titanium alignment wires in different deflection configuations. Angle Orthod 2014;84:541-7.
[22] Sarul M, Kawala B, Kawala M, Antoszewaka-Smith J. do the NiTi low and constant force levels remain stable in vivo? Eur J Orthod 2015;#7:656-64.
[23] Nikolai RJ. On optimum orthodontic force theory as applied to canine retraction. Am J Orthod 1975;68(3):290-302
[24] Alberto T., 2010, Sentalloy^R the Story of superelasticity, Department of Orthodontics, Universidad Technologica de Mexico, Maxico City, p1-9.
[25] Santoro M., Nicolay OF, Cangialosi TJ.: Pseudoelasticity and termoelasticity of nickel-titanium alloys: A clinically oriented review. Part II: Deactivation forces. Am J Orthod Dentofac. Orthop. 2001, 119,594-603.
[26] Bishara SE., Winterbottom JM., Sulieman AA., Kwan RB., Jakobsen JR. : Comparisons of the thermodynamic properties of three nickel-titanium orthodontic archwires. Angle Orthod 1995,65(2),117-122.
[27] Santoro M, Nicolay OF, Cangialosi TJ. Pseudoelastisticity and thermoelasticity of nickel-titanium alloys: a clinically oriented review. Part II: deactivation forces. Am J Orthod Dentofac Orthop 2001;119:594-603.
[28] Sifakakis I, Pandis N, Makou M, Eliades T, Bourauel C. A comparative assessment of the forces and moments generated with various maxillary incisor intrusion biomechanics. Eur J Orthod 2010;32:159-64.
[29] Sifakakis I, Pandis N, Makou M, Eliades T, Bourauel C. A comparative assessment of the forces and moments generated at the maxillary incisors between conventional and self-ligating brackets using a reverse curve of Spee NiTi archwire. Aust Orthod J 2010;26:127-33.

IOP Conf. Series: Materials Science and Engineering **265** (2017) 012014 doi:10.1088/1757-899X/265/1/012014

An Experimental Investigation of the Effects of Femtosecond Laser Helical Drilling: Influence of Process Parameters

Yancheng You[1], Yiwei Dong[1,2*], Xiaoji Li[1], Qi Zhao[1], Ertai Wang[1], Chunping Yin[1]

1. School of Aerospace Engineering, Xiamen University, 361005, Xiamen, China
2. Shenzhen Research Institute, Xiamen University, 518057, Shenzhen, China

* Corresponding author: yiweidong@xmu.edu.cn

Abstract. Femtosecond laser pulses used in micro-drilling, which allows precise and thermal-damage-free removal of material, has progressed remarkably in recent years to become an essential tool for microhole drilling. Helical drilling is the most common method for processing high-precision microholes. Compared to multi-pulse drilling or circular scanning drilling, it is more convenient to process the requested radius and needed depth hole. The mechanism of interaction between the ultra-fast laser and materials is very complex. Exploring the influence of processing parameters on the drilling process not only helps to guide the actual processing, but also helps our understanding of the mechanism. In this study, laser processing parameters for drilling microholes in three materials are investigated. The influence of processing parameters on hole drilling is analysed, and the relationship between the overlapping rate influence on drilling depth and ablation threshold is explored.

1. Introduction

Laser manufacturing has been adopted as an effective technology for producing micro-features in various materials, such as metal, silicon, and ceramic[1]. The great demand for precision holes has promoted the rapid development of laser drilling in many fields[2].For example, laser drilling has been applied in the aerospace industry, manufacturing shaped film holes in turbine blades. These film holes vary significantly in morphology and inclination angle, forming a protective cooling layer between the hot gases and external surfaces of the turbine blade[3].

Due to the complexity of the interaction regime between the femtosecond laser beam and material, there is not yet a complete theory which can describe the film cooling hole drilling process based on femtosecond laser. For the work presented in this paper, a helical drilling optic with a femtosecond-pulsed laser source is used for the drilling of microholes in stainless steel, aluminum alloy, and nickel-based alloy, respectively. The objective of the paper is to investigate the main influencing factors on quality and productivity, such as laser energy, repetition rate, and rotation speed.

2. Experiment set-up

In our case, an ultra-fast laser machine tool (KH7040A-1) from China KeHan Laser is used. The structure diagram of the machine is shown in Figure 1. The laser beam first passes through a quarter-wave plate to ensure circular polarization of the beam when reaching the workpiece, and then through a beam expander, before it is reflected by the mirror. Next, the beam passes through a beam rotation apparatus consisting of four optical wedges, which makes the beam rotate in the set helical path. The schematic diagram of the helical path can be easily observed through the SEM images of a 5s femtosecond laser pulse helical drilled hole, shown in Figure 2.

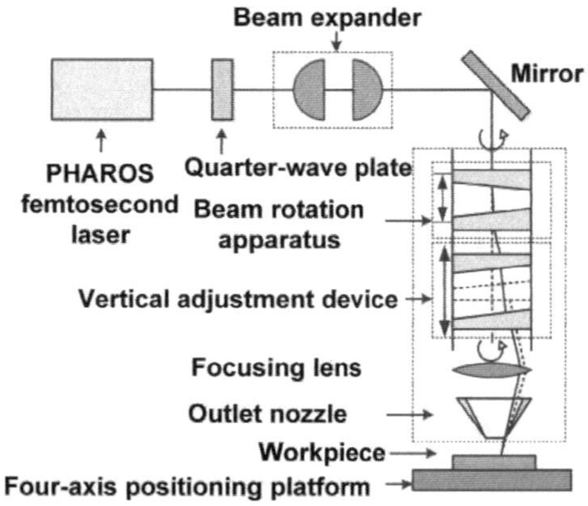

Figure 1. Schematic diagram of the femtosecond laser drilling system.

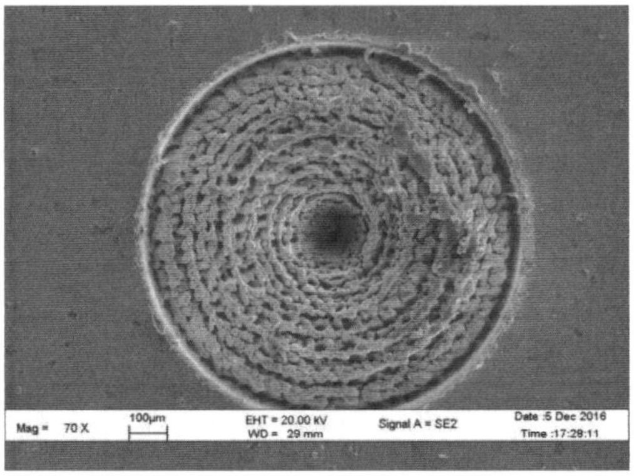

Figure 2. SEM images of a five second femtosecond laser pulse helical drilled hole.

For the experiment set-up, we chose three processing parameters: the laser repetition rate, rotation speed, and average laser power, as the variables for drilling microholes in four different materials. We measured the depth of the drilled holes by cutting the holes from the center of each using a scanning electron microscope (SEM). Furthermore, we aimed to explore the influence of the processing parameters on the processing efficiency. We set the processing parameters according to the actual hole processing situation, and set the parameter range near the common value shown in Table 1. The experiment parameters are summarized in Table 2. Based on the technician's suggestion, according to the material properties, the average power is set to 12.5W for processing on aluminum alloy and the others are set to 10W.

Table 1. Helical drilling parameters.

Repetition (kHz)	Pulse width (fs)	Spot diameter (μm)	Wavelength (nm)	Pulse average power (W)	Rotation speed (r/min)
100	250	10-15	1064	10, 12.5	2400

Table 2. Three-factors experimental parameters.

Number	Independent variable	Range	Interval	Drilling time
1	Repetition rate	60-600kHz	90	10s
2	Rotation speed	600-2400r/min	600	10s
3	Average power	3-15W	3	5s

3. Results and discussion

3.1 Laser processing parameters

First, we need to know what occurred inside the material when processing a microhole. The evolutionary process of the hole shape is shown in Figure 3. From the graph, we can divide the process into three phases. In the first phase, the ablation started from the central region and gradually spread out. In the second phase, the processing depth increased and the center point penetrated first approximately 4-6s after starting. In the third phase, the material was ablated as time went on and the hole on the back became bigger. During the processing, the center point was the deepest and quickly reached the back of the material, where the other regions were removed slowly. This is owing to the constant rotation speed and laser output power. The laser energy for each cycle is constant, despite the influence of other factors on the ablation volume. The same laser energy absorbed by the material resulted in a smaller diameter of helical path, the deeper it penetrated in this position.

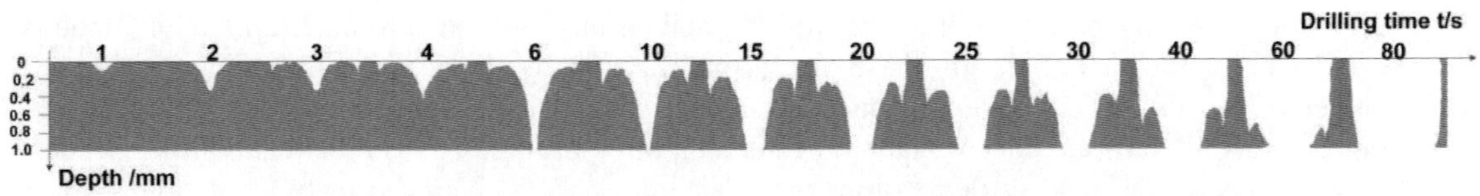

Figure 3. Temporal evolution of hole shape from the cutting section.

In order to better process microholes with high quality, it is necessary to know what effect changing the parameters will have on the processing efficiency. As shown in Figure 4, there are experimental results showing the relationship between the depth of the holes and processing parameters for the three different materials. Because the depth of each different radius is not the same, the depth of the center point is taken as the index. As can be seen from Figure 4a, the depth of ablation decreases as the repetition rate increases. This is because the single pulse energy E_p can be calculated by equation (1):

$$E_p = P \cdot f^{-1} \tag{1}$$

Where P is the average power of laser, and f is the repetition rate of laser. It is clear that an increase in repetition frequency results in a decrease in single pulse energy. From Figure 4c, we learn that the ablation depth decreases with the decrease of the single pulse energy. Thus, the increase of the repetition frequency leads to the lowering of ablation depth, thus affecting the processing efficiency.
With the increase of rotation speed ω_m, the hole depth rises. This is related to the overlapping rate, which will be discussed in the next section.
In order to better understand the influence of laser processing parameters on machining efficiency, assuming a linear relationship between parameters and depth, the linear fitting of experimental data is obtained. Table 3 shows the linear change rate of three kinds of materials. For each of the three materials, the trend is consistent, $S_E > S_f > S_\omega$, which means that the single pulse energy has the greatest influence on the hole depth, followed by the repetition frequency, and finally, the rotation rate, and a, b, and c represents the slope of the repletion rate curve, rotation speed curve, and the single pulse energy curve, respectively.

Table 3. Influence of Process Parameters on Drilled Hole Depth of Different Materials.

Material	S_f^a	S_ω^b	S_E^c
Steel304	-0.6030	0.3233	4.6163
Al5052	-0.8109	0.2846	3.9232
dd6	-0.4940	0.1859	4.9520

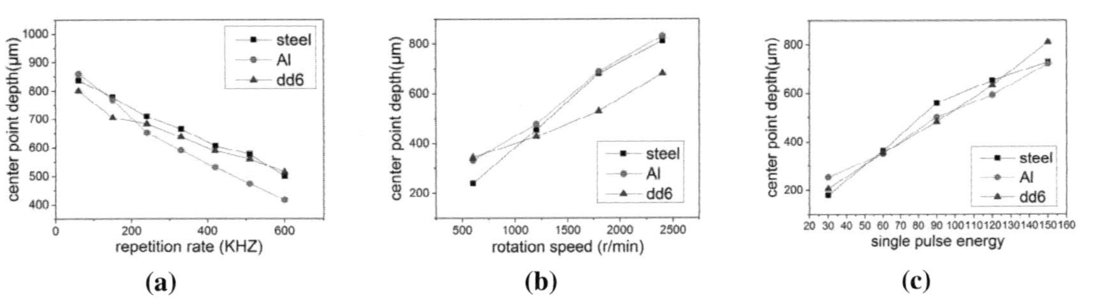

Figure 4. Correlation between removal thickness and (a) repetition rate, (b) rotation speed and (c) single pulse energy.

3.2 *Overlapping rate*

In the machining process using ultra-short laser in the helical path, we need to determine what actually changed in the laser beam by changing these processing parameters. Owing to a high repetition rate, a large number of pulses are deposited in one revolution, which results in the overlap of the pulses. According to many literatures, the overlapping rate, as a key parameter in the helical drilling process for directly describing the energy deposition, plays an important role in the drilling of high quality holes. The calculation of the overlapping rate η_{or} of the laser spot in the helical drilling is given as shown in equation (2):

$$\eta_{or} = 1 - \frac{r_h(z) \cdot \sin(2\pi \cdot \omega_m / f)}{\omega(z) + \pi \cdot \omega_m \cdot r_h(z) \cdot \tau_p} \tag{2}$$

In the formula, $r_h(z)$ is the helical radius of the circle when the laser spot is at z position, ω_m represents the rotation speed of the laser beam, $\omega(z)$ is the radius of the laser spot in z position, and τ_p is the duration of a single pulse.

To simplify the above equation to make the following calculations easier, because τ_p is 250fs, and $f >> \omega_m$, thus, it can be considered that $\pi \cdot \omega_m \cdot r_h(z) \cdot \tau_p \to 0$ and $\sin(2\pi \cdot \omega_m / f) \to 2\pi \cdot \omega_m / f$. Then, we can obtain the following formula as[4]:

$$\eta_{or} = 1 - \frac{2\pi \cdot \omega_m \cdot r_h(z)}{\omega(z) \cdot f} \tag{3}$$

Now, we have the repetition rate f and rotation speed ω_m, the maximum radius of the hole, 1mm, as the helical radius $r_h(z)$, and the radius of the spot $\omega(z)$ is 10μm. Thus, the overlapping rate can be calculated. Figure 5 depicts the depth of the center point with the overlapping rate. As shown in the figure, the depth of the microhole is inversely proportional to the overlapping rate. The slope of the curves denotes the sensitivity of the ablation depth to the overlap rate. Obviously, it is easy to conclude when discussing the influence of the overlapping rate, that it is related to the material properties. Moreover, from our experiment, the order of influence is Steel304>dd6.

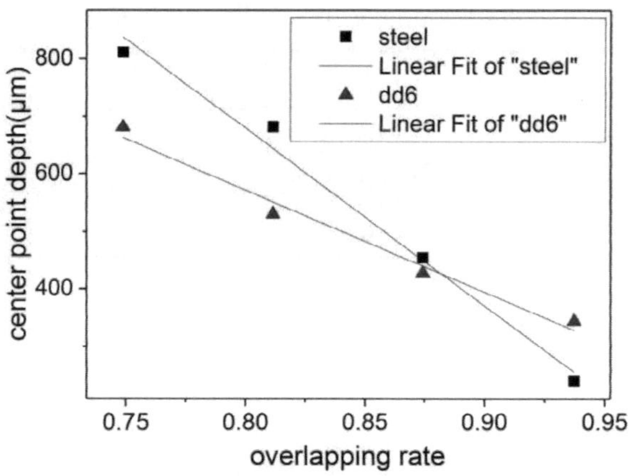

Figure 5. Correlation between overlapping rate and the depth of drilling hole.

What resulted in this phenomenon? We need to think from the source. Why can the overlapping rate affect the depth of the holes? One popular explanation is because of the energy shielding due to adsorption by vaporized particles and plasma created with ultrashort pulse laser ablation[5,6]. We assume that the greater influence of the overlapping rate is because of the energy shielding making it more difficult to reach the ablation threshold of the material, which means that the threshold is bigger, and the influence of the overlapping rate is more obvious.

As for now, the general method to get the ablation threshold of a material is using the following equation with single pulse drilling experimental data:

$$D^2 = 2\omega_0^2 \ln(-\frac{E_p}{E_{th}}) \tag{4}$$

In the formula, ω_0 is the waist radius after laser focusing.

Within the femtosecond pulse width, the material has a fixed ablation threshold[7]. There are few studies on femtosecond laser ablation thresholds, as the picosecond laser processing and femtosecond laser processing mechanisms are similar, we can learn through the investigation of the picosecond laser drilling experiment. An experimental study is presented of microhole drilling on the surface of stainless steel using a 10ps Q-switched Nd:VAN pulsed laser at two wavelengths 532 and 1064nm. From the measured data, the calculated thresholds are 0.276 J/cm^3 and 2.537 J/cm^3 [8]. Experimental results demonstrate that the ablation threshold of dd6 in the picosecond laser is 0.11 J/cm^3 [9]. Thus, we can probably draw the conclusion that the ablation threshold of stainless steel 304 is indeed bigger than the nickel-based single crystal alloys dd6s, which fits our guess.

4. Conclusion

Through the laser processing parameters in helical drilling experiment, the influence of parameters on the depth of the hole is obtained: with the increase of rotation speed, decrease of laser repetition frequency, and increase of the single pulse energy, the depth of the hole is increasing, which denotes that the drilling efficiency is higher, within the range of values taken in the paper. From the shape evolution of the hole inner shape, which can be obtained by cutting holes from the center, the drilling process can be divided into three phases. The depth decreases with the larger radius, and the center point will penetrate first. The overlapping rate has a great impact on the drilling process because of the energy absorption by vaporized particles and plasma. The influence of the overlapping rate on the drilling process is related to the material, and likely to have a relationship with the material ablation threshold. The more difficult to ablate, the bigger the threshold, and the impact of overlapping is more obvious.

Acknowledgments

The authors are grateful for the financial support provided by the National Natural Science Foundation of China (51705440), the PhD Start-up Fund of the Natural Science Foundation of Guangdong Province (2014A030310004), the Major Science and Technology Project between University--Industry Cooperation in Fujian Province (2015H6023), and the China Scholarship Council. In addition, the authors would like to thank China KeHan Laser (Fujian) Co.,Ltd for providing the equipment used in this study.

References

[1] Cheng J, Perrie W, Sharp M, Edwardson SP, Semaltianos NG, Dearden G and Watkins KG 2009 *Appl. Phys. A*: **95(3)** 739–746
[2] Sugioka K, Meunier M and Piqué A 2010 *Springer.*
[3] Bunker RS 2005 *Journal of Heat Trans* **127(4)**: 441.
[4] Gillnera A 2016 *InProc. of SPIE* **9741** 974106–1
[5] Ancona A, Röser F, Rademaker K, Limpert J, Nolte S and Tünnermann A 2008 *Opt. Express* **16(12)**: 8958–68
[6] Park KW and Na SJ 2010 *Appl. Surf. Sci* **256(8)**: 2392–9
[7] Kurt TH, Ferdinand VA, Ostendorf A, Kamlage G, Nolte S 1999 *Int J Elec Mach* **4** 1–6
[8] Zhao W, Wang W, Jiang G, Li B and Mei X 2015 *Int J Adv Manuf Technol* **1**:81
[9] Ji L, Zhang XB, Zhang W, Sun RF and Han JG 2014 *Appl Laser* **34**, 551–556

IOP Conf. Series: Materials Science and Engineering **265** (2017) 012015 doi:10.1088/1757-899X/265/1/012015

Microstructure and tensile properties of Ti-6Al-4V alloys manufactured by selective laser melting with optimized processing parameters

L Wang, C Ma, J Huang, H Y Ding and M Q Chu

Shanghai Aircraft Manufacturing Co., Ltd, Shanghai 201324, China

wanglei42@comac.cc

Abstract. Selective laser melting (SLM) is a precise additive manufacturing process that the metallic powders without binder are melted layer by layer to complex components using a high bright fiber laser. In the paper, Ti-6Al-4V alloy was fabricated by SLM and its microstructure and mechanical properties were investigated in order to evaluate the SLM process. The results show that the microstructure exists anisotropy between the horizontal and vertical section due to the occurrence of epitaxial growth, and the former microstructure seems equal-axis and the latter is column. Moreover, there is little difference in tensile test between the horizontal and vertical sections. Furthermore, the tensile properties of fabricated Ti-6Al-4V alloy by SLM are higher than the forged standard ones. However, the fatigue results show that there are some scatters, which need further investigation to define the fatigue initiation.

1. Introduction

Additive Manufacturing (AM) is a process in which parts are built up by progressive consolidation of raw materials, such as powder or wire, in a layer-by-layer fashion [1–4]. This is a different approach to traditional fabrication methods such as machining from block or plate, where the final part geometry is produced by subtracting or removing material. All additive layer manufacturing technologies share the layer additive approach: a three-dimensional CAD model is sliced into thin layers and the particular AM process builds the physical part layer by layer. With this additive approach, parts of greater complexity can be economically produced. Manufacturing near-net-shape components in a layer-by-layer fashion offers a great potential for time and cost savings in comparison to conventional manufacturing technologies. Especially aerospace components that are machined from costly wrought material at a low fly-to-buy ratio represent interesting applications [5-7]. Recently, there exist some reports dealing with microstructures of Ti-6Al-4V alloy fabricated by AM processes, but none of them is related to evaluating the SLM process in microstructure and mechanical properties in different locations in order to satisfy the industrial requirements [8-10]. In the present paper, Ti-6Al-4V alloy was deposited by commercial SLM processes, and its microstructure and tensile properties were investigated for aeronautical industry application. The material investigated is the titanium alloy Ti-6Al-4V, the most commonly produced titanium alloy and one of the most common aerospace alloys.

2. Experiment

Ti-6Al-4V alloy powder was provided by Falcon Rapid Fabrication Technology Corporation. The powder was produced by the gas atomization process and was spherical with a maximum particle size 53 μm (98.7% less than 53 um). Ti-6Al-4V alloy samples were fabricated on an EOS M280 machine including a laser unit delivering a continuous single mode laser power of 400 W which produces a laser beam with a wavelength of 1070 nm and an intensity distribution of Gaussian. The laser spot diameter was 100~500 μm and the maximum scanning speed was 7 m/s. In addition, the layer thickness can be selected between 20~100 μm and the deposition was carried out on a 30 mm thick Ti-6Al-4V alloy plate. During the SLM process, the processing chamber will be filled with argon in order to maintain the oxygen level during the process. The schematic principle of SLM process is shown in Figure 1.

The horizontal and vertical section microstructures of the samples were examined using a Leica optical microscope. The chemical composition was carried out with the Oxford INCA energy dispersive X-ray (EDX) microanalysis software. The element concentration conforms to AMS4998 regulation. Two post heat treatments were applied. Stress relieving at 800 ℃ for 2 hours followed by furnace cooling (FC) to reduce residual stresses without substantial change to microstructure.

The sample location was shown in Figure 2 and the room temperature tensile property was tested according to ASTM E8/E8M. Samples were taken from middle point and four corners of the base, the specimens size were Φ10×70 mm for vertical samples and 10×10×70 mm for horizontal samples. A schematic drawing of the test pieces is shown in Figure 2.

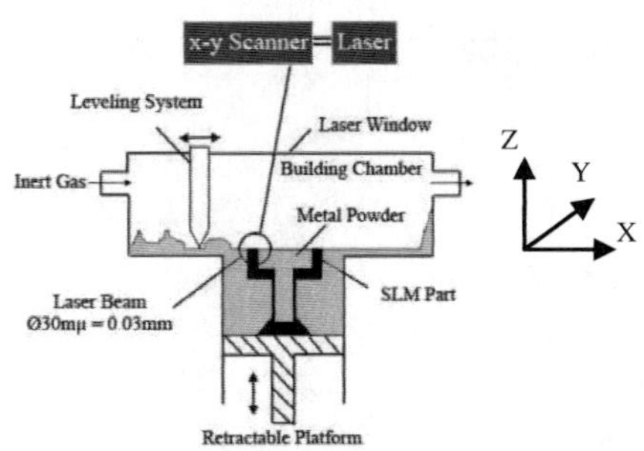

Figure 1. Schematic diagram of SLM process.

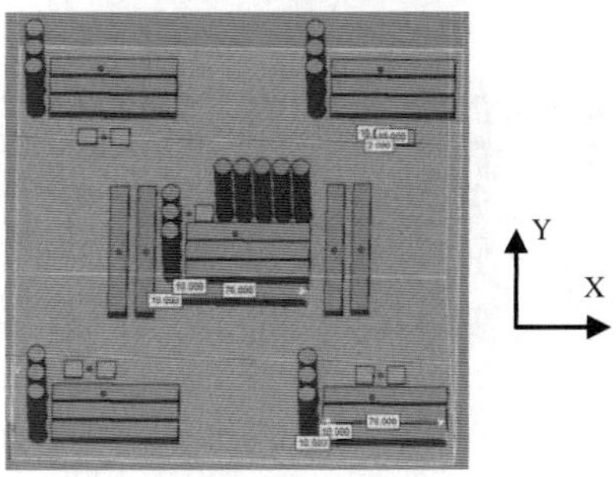

Figure 2. Drawing of the tensile testing bars.

3. Results and Discussions

3.1. Microstructure

Figure 3 shows the microstructure features of Ti-6Al-4V alloy fabricated by SLM in the X-Z and X-Y sectional directions. The microstructure shows morphology as macrostructure of columnar prior-grains that are growing epitaxially across many layers vertically in Figure 3(b). This type of microstructure is a typical of additive manufacturing processes, which involve partial re-melting of the previous layers. Indeed, when the laser beam scans the layer of powder, it remelts the top of the columnar grains, which results in the epitaxial growth of the strongly textured grains. Moreover, the time the melting pool remains liquid before a complete solidification in the β domain, depending essentially on the solidification rate and the thermal gradient, acts on the epitaxial growth and especially the number of nuclei formed. When a layer of the former microstructure seems equal-axis and the latter one is column. It seems to mean that this unidirectional grain growth may also influence the anisotropy of properties of the SLM as-built alloy. However, the electron back scatter diffraction technique shows that there is no obvious texture in both sections as shown in Figure 4.

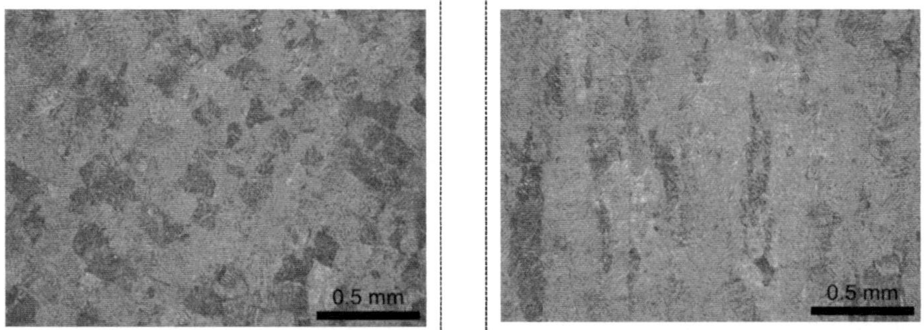

(a) Microstructure of Ti-6Al-4V alloy fabricated by SLM: cross-section through columnar grains of the prior-β phase in the X-Y plane.	(b) Longitudinal section through columnar grains of the prior-β phase in the X-Z.

Figure 3. Microstructure features of Ti-6Al-4V fabricated by SLM in the X-Z and X-Y sectional directions.

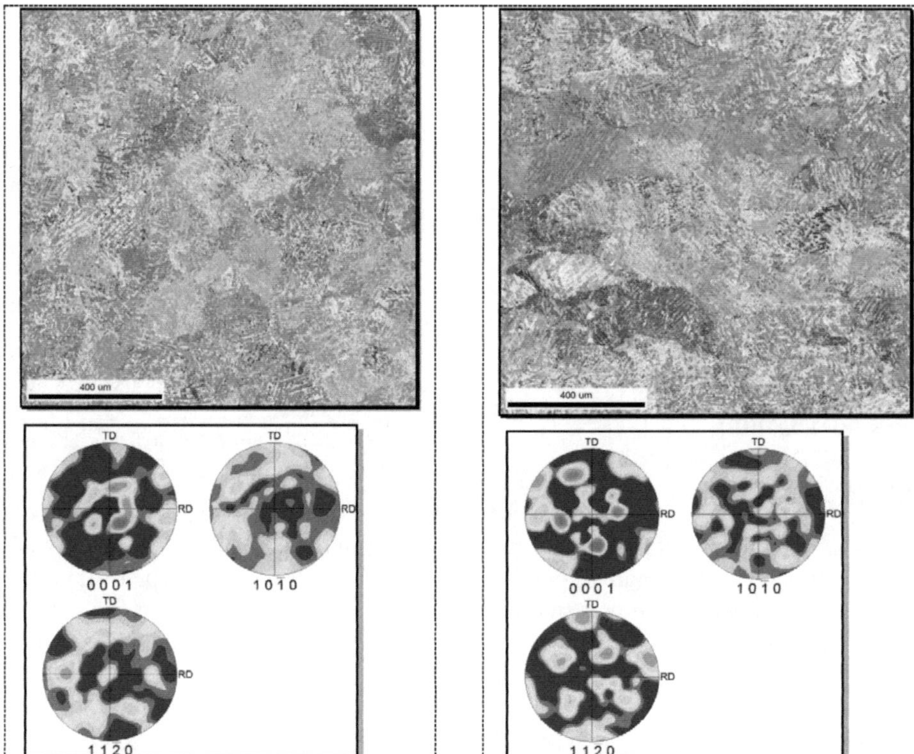

(a) cross-section through columnar grains of the prior-β phase in the X-Y plane.	(b) longitudinal section through columnar grains of the prior-β phase in the X-Z plane.

Figure 4. EBSD of Ti-6Al-4V alloy fabricated by SLM.

3.2. Tensile properties

Table 1 and Table 2 show that the tensile properties of Ti-6Al-4V alloy fabricated by SLM in the X-Y and X-Z sections at room temperature. It can be seen that the SLM tensile strength is similar in both transverse and longitudinal sections, no matter the sample location is in the middle part or in the corner position during the processing, which is higher than the as-forged standard one (the tensile properties of forged Ti-6Al-4V alloy is yield strength of forged 828MPa, ultimate tensile strength 895MPa and elongation 10%). However, the fatigue test results, as shown in Figure 5, show that there are some scatters between the two sections in spite of the limited data, due to some defects occur during the processing or the powder.

Table 1. Tensile properties of Ti-6Al-4V alloy fabricated by SLM in X-Y direction at different location of the base.

No.	UTS (MPa)	Standard Deviation (MPa)	CV (%)	YS (MPa)	Standard Deviation (MPa)	CV (%)	E (%)	Standard Deviation (%)	CV (%)
c1h	1037	5.77	0.56	980	5.00	0.51	15.5	0.87	5.58
c2h	1040	0.00	0.00	982	2.89	0.29	15.83	1.26	7.95
c3h	1067	15.28	1.43	1018	20.21	1.98	16.83	0.76	4.54
c4h	1053	5.77	0.55	988	7.64	0.77	14.17	1.04	7.35
mh	1057	11.55	1.09	997	12.58	1.26	16.17	1.04	6.44

Table 2. Tensile properties of Ti-6Al-4V alloy fabricated by SLM in X-Z direction at different location of the base.

No.	UTS (MPa)	Standard Deviation (MPa)	CV (%)	YS (MPa)	Standard Deviation (MPa)	CV (%)	E (%)	Standard Deviation (%)	CV (%)
c1h	1033	11.55	1.48	977	14.43	1.12	15.83	1.04	6.57
c2h	1037	5.77	1.64	983	16.07	0.56	15.83	0.76	4.82
c3h	1037	5.77	1.30	967	12.58	0.56	16.17	1.76	10.86
c4h	1047	5.77	1.47	982	14.43	0.55	15.83	0.29	1.82
mh	1067	5.77	1.14	1013	11.55	0.54	13.83	0.58	4.17

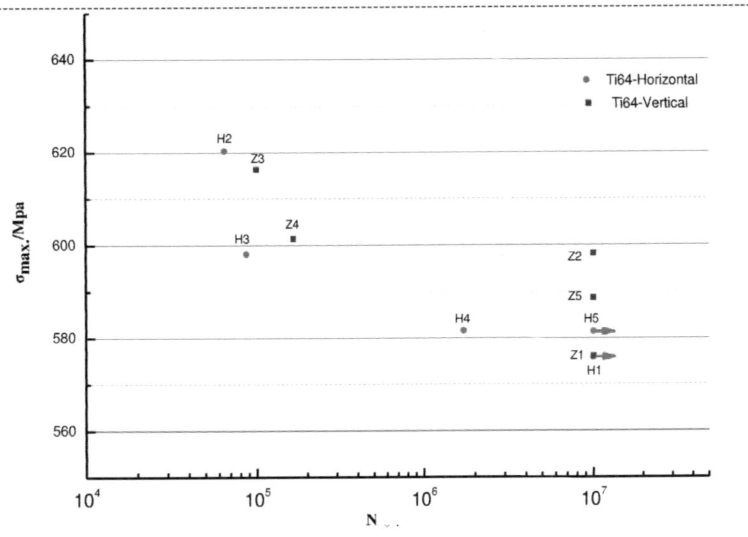

Figure 5. Fatigue test results of the horizontal and vertical sections.

4. Summary

(1) The microstructure of fabricated Ti-6Al-4V alloy by SLM shows anisotropy between the X-Z section and X-Y section. The former microstructure is column and the latter seems equal-axis owing to epitaxial growth. In addition, there is no obvious texture occur between the X-Z section and X-Y section.

(2) The tensile strength and ductility of fabricated Ti-6Al-4V alloy by SLM are quite uniform on the different process position, showing the higher value than that of the forged standard ones.

(3) The fatigue test results of fabricated Ti-6Al-4V alloy by SLM show some scattering due to some defects occur during the processing or the powder.

5. Acknowledgments

The present work was funded by the Foundation of Shanghai Science and Technology Commission, China (No.15520701400) and Commercial Aircraft Corporation of China, Ltd.

6. References

[1] Kumar S 2014 *Comprehensive Materials Processing* **10** 93
[2] W D Huang and X Lin 2010 *Materials China* **29** 12
[3] L E Murr and S M Gaytan 2014 *Comprehensive Materials Processing* **10** 135
[4] F D Wang, S Williams, P Colegrove and A A Antonysamy 2013 *Metallurgical and Materials Transactions A* **44** 968
[5] M Q Chu, L Wang, H Y Ding and Z G Sun 2015 *Applied Mechnicas and Materials* **798** 457
[6] L Xin and W D Huang 2015 *Materials China* **34** 684
[7] H M Wang 2014 *Acta Aeronautica et Astronautica Sinica* **35** 2690
[8] P Edwards and M Ramulu 2014 *Materials Science and Engineering A* **598** 327
[9] K N Amoto, S M Gaytan and L E Murr 2012 *Acta Mater* **60** 2229
[10] P Edwards and Ramulu M 2014 *Mater Sci Eng A* **598** 327

IOP Conf. Series: Materials Science and Engineering **265** (2017) 012016 doi:10.1088/1757-899X/265/1/012016

Transverse bending deformation analysis of 3D surface part in continuous roll forming process

M Wang[1*] , S C Yang[1], G L Lu[2] and J H Cao[3]

[1]School of Engineering, Changchun Normal University, 677 Changji North Road, Changchun, P. R. China, 130032
[2]Key Laboratory for Bionic Engineering (Ministry of Education), Jilin University (Nanling Campus), 5988 Renmin Street, Changchun, P. R. China, 130025
[3]School of Mechanical and Electrical Engineering, Xinxiang University, 191 Jinsui Street, Xinxiang, P. R. China, 453003

wangmi3214@126.com

Abstract. This paper introduces a newly developed technique called continuous roll forming process which is applied to manufacture 3D surface part with various shapes, especially for doubly curved parts. In continuous roll forming process, the twin bendable rolls are placed to form a nonuniform roll gap. The bendable rolls rotate in opposite directions around the bend axis with equal angular velocity. The sheet metal is bitten into the roll gap by friction force and compressed nonuniformly across its width. At the present, how to precisely predict transverse bending deformation of 3D surface part has become a critical issue for guiding practical forming in the continuous roll forming process. This work focuses on analysis of the transverse bending deformation characteristics in continuous roll forming process.

1. Introduction

Roll bending process is widely used to manufacture cylindrical or conical sections with constant cross section[1]. However, the requirements for products in production practice tend to be personalized and varied, needing more completed surface part. Flexible roll forming is a new way to manufacturing various 3D surface parts with complex cross section. Discrete-roll forming was developed by Shim et al by using the multiple roll sets to form doubly curved plates in whole region with only one pass[2, 3]. Li et al proposed continuous flexible forming process composed of an up and two down flexible rolls, and the configuration of three flexible rolls are adjusted by multiple points on the basis of the shape of the target part[4]. Continuous roll forming process explored by Cai et al is a novel roll bending technology to translate a flat metal plate into 3D surface part.[5, 6] As shown in figure 1, the continuous roll forming device is composed of an up and a down bendable rolls, and the configurations of the bendable rolls can be controlled in accordance with the desired shape of the formed part. The sheet metal is deformed in the local region between the up and down bendable roll, and with the bendable roll rotation with the bend axis, and the sheet metal is moved by the friction between and continuously deformed line by line.

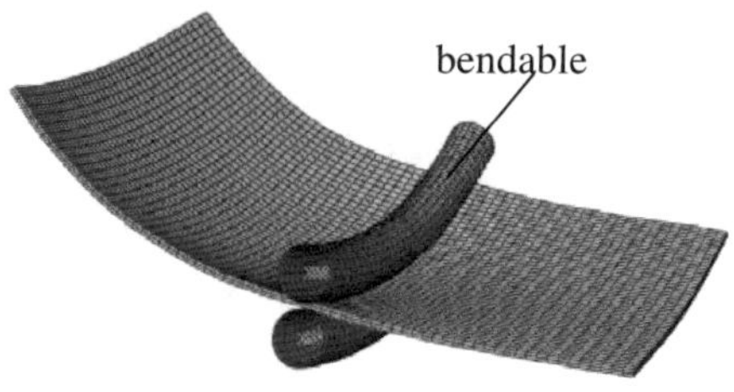

Figure 1. Diagram of geometric feature of deformation zoon

2. Transverse bending deformation in continuous roll forming process

In order to understand the transverse bending process of the sheet metal, the radius of the transverse curvature of cross section at different positions is analyzed. As shown in figure 2, the transverse bending deformation at seven positions (d_1, d_2, d_3, d_4, d_5, d_6, d_7) along the direction of motion are extracted and shown. The radius of the transverse curvature of cross section at d_7 is 560.1 mm, and the radius of the transverse curvature is decreased from d_7 to d_1 and reaches minimum at d_1 with the value 251.6 mm. It can be found from the changing regularity of the radius of the transverse curvature of cross section, the sheet metal has been bent in a certain distance from the roll gap, and the radius of the transverse curvature of the sheet metal is becoming more close to the radius of the transverse curvature of the bendable roll when closing to the deformation zoon. It can be assumed that the sheet metal deforms gradually in the process of approaching the bendable roll, and the transverse bending deformation is finally completed in the contact deformation zone. Figure 3 shows the variation of transverse curvature of sheet metal, the ratio of the distance to the roll gap to the sheet metal width is abscissa, and the transverse curvature at corresponding position is ordinate. For different positions on the sheet metal, at more distant positions from the roll gap, the less transverse curvatures are found, that is to say, the transverse deformation of the sheet metal is produced gradually.

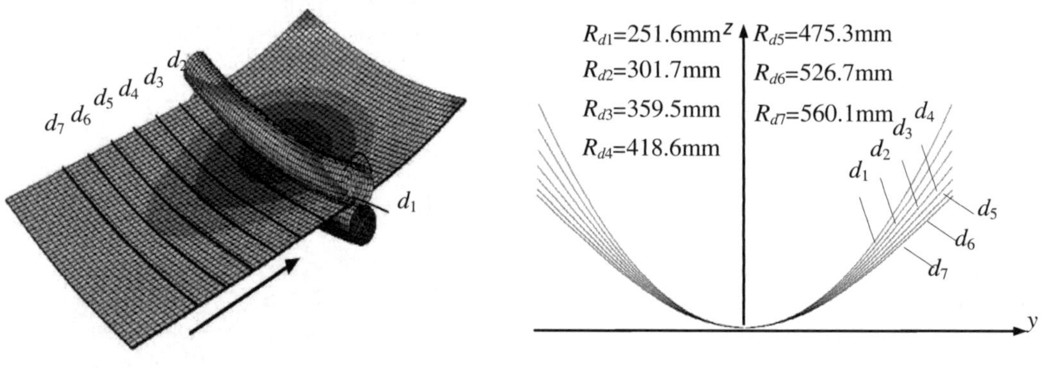

(a) Simulation process (b) Radius of the transverse curvature of cross section

Figure 2 Profile curves and radii of curvatures of cross-section

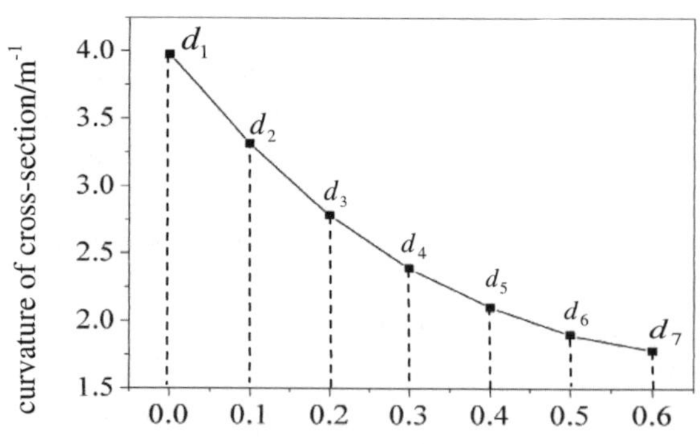

Figure 3. Variation of curvature of cross-section `

3. Numerical simulations

In the continuous roll forming process, the sheet metal deformed in the deformation zoon in a very short time, and the forming process is affected by many factors, so it is difficult to observe the deformation process through experiments. The numerical method was employed to study the deformation process in the continuous roll forming process. Based on the numerical simulation, visualization results of the deformation process can be obtained. Equivalent stress of the sheet metal in continuous roll forming is shown in figure 4.

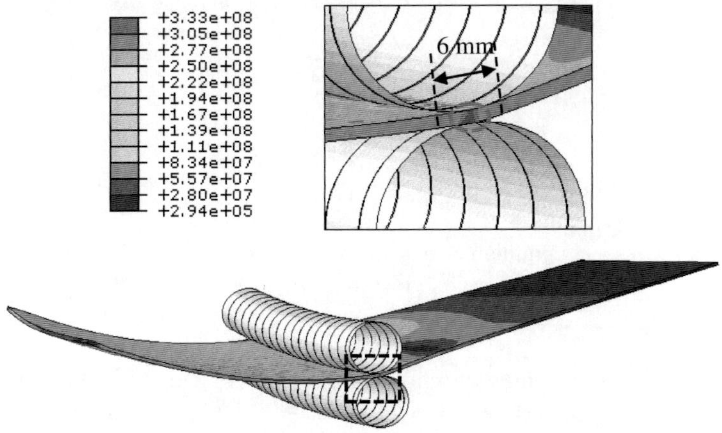

Figure 4. Equivalent stress in continuous roll forming

According to the above analysis, the stress condition of the sheet metal in different positions of the deformation zoon is varied. Figure 5 shows the diagram of geometric feature of deformation zoon, it can be observed that the equivalent stress of the sheet metal between the up and down bendable rolls is large, and the maximum value of 333 PMa appears in the deformation zone. Since the equivalent stress varies greatly near the deformation zoon, the stress state in the area near the deformation zone with width 6 mm is analyzed. The stress state in deformation zone is complex and divided into several zones. Figure 5 gives the cross section (*EE′*, *AA′*, *DD′*) in deformation zone and the longitudinal fibers of the formed parts. *AA′* is in the middle of the deformation zone, *EE′* and *DD′* are the cross sections with the distance of 3mm from the left and right sides of the *AA′*. The longitudinal fiber l_1, l_2, l_3, l_4, l_5 are uniformly distributed between the middle and the edge.

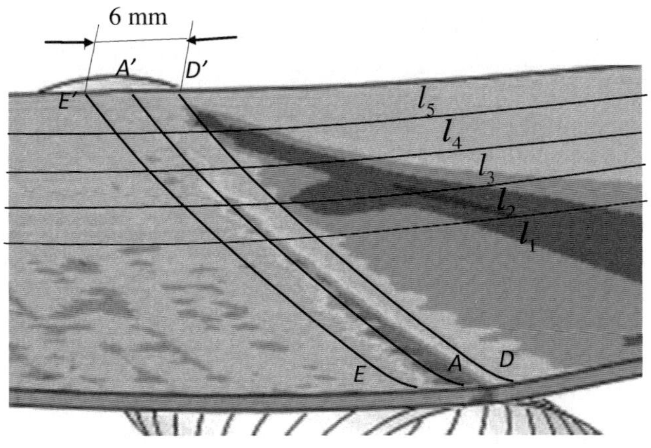

Figure 5. Diagram of geometric feature of deformation zoon

Figure 6 shows the equivalent stress distribution and minimum principal stress distribution. It can be found that the equivalent stress is large in the middle (*AA′*) and small on both sides (*EE′* and *DD′*), and from l_5 to l_1 the equivalent stress are gradually decreased. Contrary to the equivalent stress, the minimum principal is small in the middle (*AA'*) and large on both sides (*EE'* and *DD'*), and the minimum principal increases gradually from l_5 to l_1.

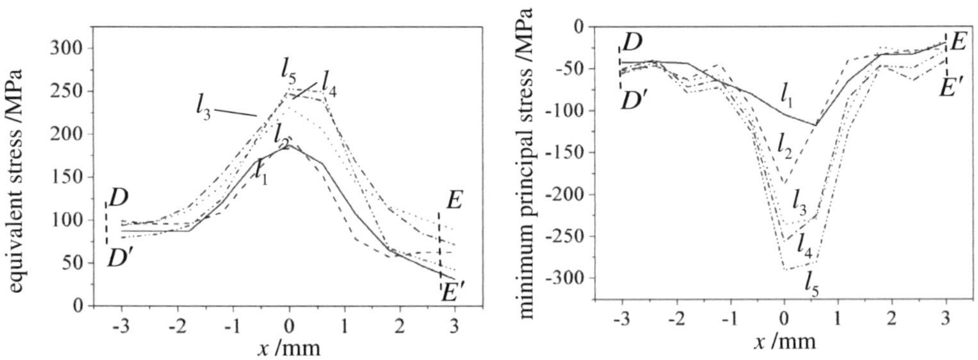

Figure 6. Equivalent stress and minimum principal stress in deformation zoon of negative Gaussian curvature

4. Conclusions

Continuous roll forming is a novel forming technology for manufacturing 3D surface part based on bendable rolls. This paper focuses on the transverse deformation of sheet metal, the transverse deformation of sheet metal at different position is studied through simulation and the stress distribution of the sheet metal in the forming process is analyzed. The numerical simulation results show that the sheet metal has been bent in a certain distance from the roll gap, and the transverse curvature of the sheet metal is becoming more close to the transverse curvature of the bendable roll when closing to the roll gap. For manufacturing a Negative Gaussian curvature part, the equivalent stress is large in the middle and small on both sides. Contrary to the equivalent stress, the minimum principal is small in the middle and large on both sides.

Acknowledgements

The authors would like to acknowledge the financial support provided by the National Natural Science Foundation of China (Grant No. 51605187, Grant No. 51605188, Grant No. 51575231, Grant No. 51375204), and Jilin Provincial Science & Technology Department (20170520088JH), Supported by China Post doctoral Science Foundation (Grant No. 2016M601382).

References

[1] Abeyrathna B, Rolfe B, Hodgson P, Weiss M. Local deformation in roll forming. International Journal of Advanced Manufacturing Technology. 2016:1-11.
[2] Shim DS, Yang DY, Han MS, Chung SW, Kim KH, Roh HJ. Experimental study on manufacturing doubly curved plates using incremental rolling process. Advanced Technology of Plasticity. 2008.
[3] Shim DS, Yang DY. Multi-stage forming using optimized preform in the line array roll set process and its industrial application. International Journal of Precision Engineering & Manufacturing. 2014;15:2085-92.
[4] Cai ZY, Sui Z, Cai FX, Liu L. Continuous flexible roll forming for three-dimensional surface part and the forming process control. International Journal of Advanced Manufacturing Technology. 2012;66:393-400.
[5] Wang M, Cai ZY, Li LL, Li MZ. Longitudinal bending deformation analysis of sheet metal in continuous roll forming process. International Journal of Advanced Manufacturing Technology. 2015;80:467-76.
[6] Cai ZY, Wang M, Li MZ. Study on the continuous roll forming process of swept surface sheet metal part. Journal of Materials Processing Technology. 2014;214:1820-7.

IOP Conf. Series: Materials Science and Engineering **265** (2017) 012017 doi:10.1088/1757-899X/265/1/012017

Thermal modeling and analysis of thin-walled structures in micro milling

J F Zhang*, Y H Ma , C Feng , W Tang and S Wang
College of Mechanical and Electronic Engineering, Shandong University of Science and Technology, Qingdao; 266590, China

zhangjf@sdust.edu.cn

Abstract: The numerical analytical model has been developed to predict the thermal effect with respect to thin walled structures by micro-milling. In order to investigate the temperature distribution around micro-edge of cutter, it is necessary to considering the friction power, the shearing power, the shear area between the tool micro-edge and materials. Due to the micro-cutting area is more difficult to be measured accurately, the minimum chip thickness as one of critical factors is also introduced. Finite element-based simulation was employed by the Advantedge, which was determined from the machining of Ti-6Al-4V over a range of the uncut chip thicknesses. Results from the proposed model have been successfully accounted for the effects of thermal softening for material.

1. Introduction

Thin-walled structures of highly accurate and miniaturized components with complex shape are becoming increasingly important in diverse field, such as aviation aerospace, biomedicine instruments, and communication systems [1].The cutting temperature, is one of the critical factors which affect the precision of parts, residual stresses, the part distortions, tool wear and so on. FEM based process simulation has been utilized to study machining processes for several years. The main advantage of process simulations remains on the fact that various variables in the machining process including those very difficult to detect by experiments such as temperature, strain, strain rate, stress etc. can be predicted.

Several groups of researchers have been working on the temperature modeling for machining processes. Usuiet et. al. [2] presented a numerical method to calculate the cutting temperature by using the predicted quantities through the energy method. The predicted temperature distribution on the tool face was found to be in good agreement with the obtained experimentally, with regard to the shape of the distribution, the maximum temperature and its location is cutting edge. However, the calculated temperature was lower than the observed one near the cutting edge and the chip leaving point. D. Ulutan, I. Lazoglu and C. Dinc [3] proposed a finite difference method-based model and determined the three-dimensional temperature filed predictions. Lazoglu et al.[4-6] developed various temperature models for different machining operations. Shear energy, friction energy, heat balance, wear et al. were all considered. Thanongsak Thepsonthi and Tu˘grul Özel [7] developed a 3-D FE modeling and simulations of micro milling of Ti-6Al-4V alloy and validated with experiments on chip flow and tool wear. In addition, temperature distribution was also studied at different edge radius size. However, there are still some research problems remain unsolved about FE process simulation such as accuracy of 2-D FE and the influence of uncut chip thickness.

Here in this paper, power model and 2-D finite element modeling are applied for the study temperatures in micro milling. These developed models were used to study the influence of increasing uncut chip thickness of micro milling.

2. Power model

In order to study temperature of micro-milling, a mechanical model of the micro milling process is required for the estimation of heat generation due to the shear friction in the primary and secondary zone. Cutting force and velocities acting on a differential cutting edge along the tool are presented in figure 1. Heat generation in the primary zone due to shearing and friction power in the second deformation zone can be determined as:

$$\begin{cases} P_s = F_s v_s = \dfrac{\tau A v \cos\alpha_e}{\cos(\varphi-\alpha_e)\sin\varphi} \\ P_f = F_f v_f = \dfrac{\tau A v \sin\beta}{\cos(\varphi-\alpha_e)\cos(\varphi+\beta-\alpha_e)} \end{cases} \quad (1)$$

Where P_s, P_f, F_s, F_f, v, τ, β, φ and α_e and A represent the shearing power, friction power, shear force on shear plane, friction force on rake plane, cutting velocity, average shear flow stress, normal friction angle, normal shear angle, and normal rake angle and shear area, respectively.

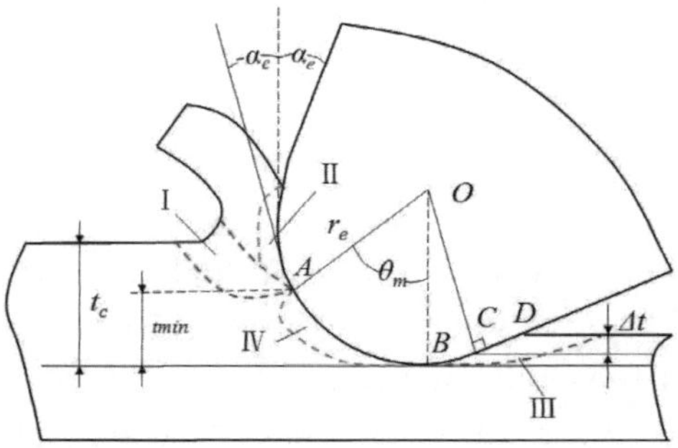

Figure 1. Force and velocity components acting on cutting edge element

In order to calculate the area of the shear plane [8-9], built the thermo-mechanical model and researched the uncut chip thickness. In micro-milling process, cutting edge can no longer be considered sharpen, cutting edge radius is comparable in size to the uncut chip thickness. So the case of $t_c \leq r_e$ is just only studied.

The engaged radius of round nose can be divided in two parts defined, respectively, by the angle θ and π-2ψ, see figure 2(*a*) with:

$$\begin{cases} \theta = \cos^{-1}(f_z/2r_e) - sin^{-1}(r_e - t_c/r_e) \\ \psi = cos^{-1}(f_z/2r_e) \end{cases} \quad (2)$$

The angle θ *is* subdivided into *N* cutting edge elements characterized by the incremental angle $\Delta\theta = \theta/N$ and by the index *j* such as $1 \leq j \leq N$. Figure 2(*b*) gives a treatment of discretization with *N*=3.

As shown in figure 2(*b*), the cutting edge element *j* can be defined by $\Delta\theta$, w_r^j, β_j, d_j, w_j^r

$$\begin{cases} \Delta\theta = \dfrac{\theta}{N}, d_j = w_r^j sin\left(\dfrac{\pi}{2} - \beta_j - \dfrac{\Delta\theta}{2}\right) \\ w_r^j = 2r_e sin\dfrac{\Delta\theta}{2},\ \beta_j = \cos^{-1}\left(\dfrac{f_z}{2r_e}\right) - \theta + (j-1)\Delta\theta \end{cases} \quad 1 \le j \le N \tag{3}$$

Where, β_j, w_r^j and d_j represent the angle position, width of cut and depth of the cut corresponding to the cutting edge element *j*.

In figure 2(*b*), the area of shear associated to the *j*th cutting edge element is defined as

$$A_j = \begin{cases} f_z t_{c_j} & 0 \le j \le N \\ \dfrac{r_e^2}{2}\left[\pi - 2cos^{-1}\left(\dfrac{f_z}{2r_e}\right) - sin\left(2cos^{-1}\left(\dfrac{f_z}{2r_e}\right)\right)\right] & j = N+1 \end{cases} \tag{4}$$

By considering equation (3) and equation (4) and after some algebraic manipulation the following expressions are obtained.

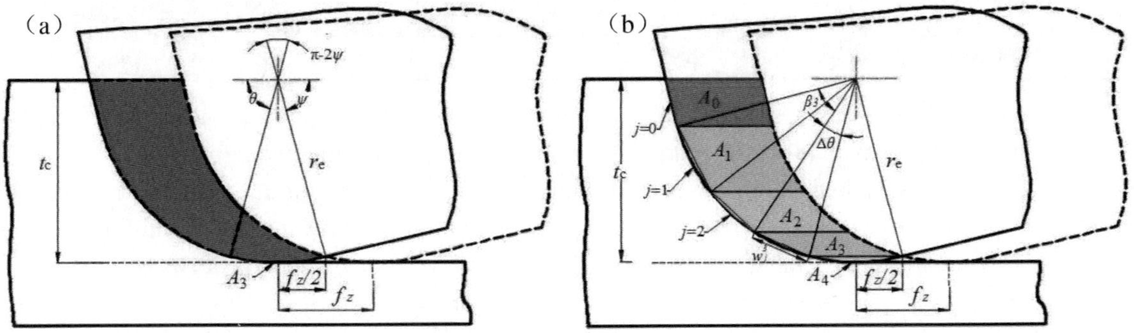

(*a*) The undiscretized model (*b*) The model of discretization

Figure 2. Subdivision of the rounded nose and the area (*N*=3 in the figure)

$$A_j = \begin{cases} 2r_e f_z sin\dfrac{\Delta\theta}{2} sin\left(\dfrac{\pi}{2} - \beta_j - \dfrac{\Delta\theta}{2}\right) & 0 \le j \le N \\ r_e^2\left[\dfrac{\pi}{2} - cos^{-1}\left(\dfrac{f_z}{2r_e}\right) - \dfrac{f_z}{2}\left(r_e^2 - \dfrac{f_z^2}{4}\right)^{1/2}\right] & j = N+1 \end{cases} \tag{5}$$

From figure 2, by using equation (5) can get

$$A = r_e^2\left[\frac{\pi}{2} - arccos\left(\frac{f_z}{2r_e}\right)\right] + \frac{f_z}{2}\left(r_e^2 - \frac{f_z^2}{4}\right)^{1/2} + f(t_c - r_e) \tag{6}$$

So the total power, which is the sum of shear and friction power, can be written as $P = P_s + P_f = cA$. Where *c* is a constant, the value is given as follow

$$c = \frac{\tau v}{cos(\varphi - \alpha_e)}\left(\frac{sin\beta}{cos(\varphi + \beta - \alpha_e)} + \frac{cos\alpha_e}{sin\beta}\right) \tag{7}$$

In this paper, we supposed $P_0 = P/c = A$.

3. 2-D finite element modeling

To construct the 2-D FE model for micro up milling of Ti-6Al-4V alloy, AdvantEdge was utilized. AdvantEdge does not need to define the separation line, chip and workpiece separation criteria and assume the chip shape in advance. So it can better avoid the grid re-division and distortion, which making the workpiece and chip to maintained a good contact state, the calculation easy to converge and promised the accuracy and efficiency [10].

AdvantEdge adopted Power Law constitutive model. It was defined by three multiplicative terms namely strain hardening, rate sensitivity and thermal softening as shown in equation (8)

$$\sigma\left(\varepsilon^{p},\dot{\varepsilon},T\right)=g\left(\varepsilon^{p}\right)\times\Gamma\left(\dot{\varepsilon}\right)\times\Theta\left(T\right) \tag{8}$$

Where, $g(\varepsilon^{p})$ is strain hardening, $\Gamma(\dot{\varepsilon})$ is strain rate sensitivity and $\Theta(T)$ is thermal softening.

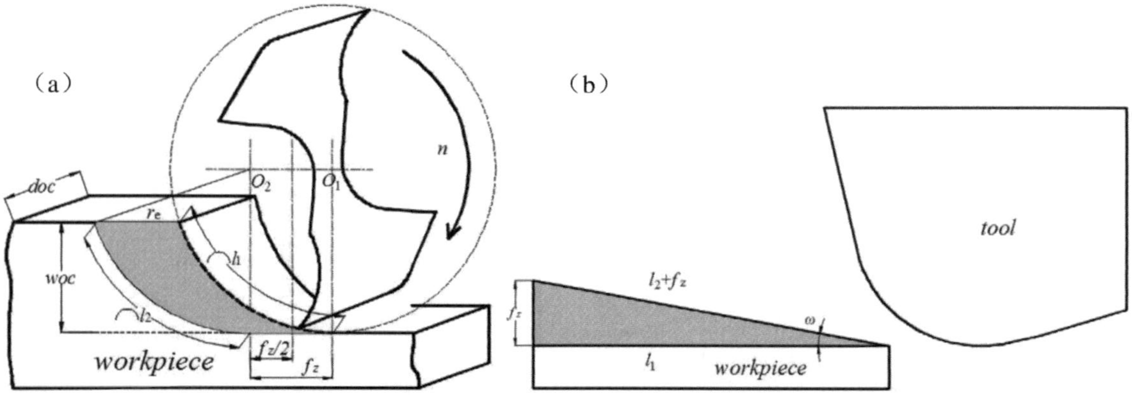

(*a*) 3-D finite element simulation (*b*) 2-D finite element simulation

Figure 3. FE model for half-immersion up micro milling schematic diagram

The figure 3 shows a schematic diagram of the milling model. From the figure 3(a), it can be noticed that the cutting quantity is indicated by shading part. Unlike other software for up milling, when converting 3-D to 2-D finite element simulation, AdvantEdge have as realistic a model as possible.

In the figure 3(b), f_t, l_1, l_2 represent the feed, arc length corresponding to the angle of rotation δ, respectively. N is the number of teeth. The relation between them can be written as:

$$l_1 = l_2 = \frac{\pi r_e}{180}\delta \,, f_t = \frac{f_z N}{2\pi}\delta \tag{9}$$

Equation (9) indicated that at the same time, there is different feed when δ is changing. However, every project actually has same feed while t_c is changing. Big t_c just took more time to complete the project. This is agreement with the process. Therefore the model, we built in AdvantEdge, is true.

In the simulation, the tool diameter is 0.5mm with a rake angle of -5° and the edge radius of 3μm. The configurations of the simulation cutting parameters are spindle speed（n）of 20000rpm, cutting depth（a_p）of 0.1mm, feed per tooth（f_z）of 5μm, and uncut chip thickness（t_c）of 1.5μm, 2.0μm, 2.5μm and 3.0μm.

4. Results

The simulation for total power of the workpiece as shown in figure 4, which is the sum of friction and shearing power. Equation (1) indicated that power is increased with the increasing shear area. That is to say, taking into equation (6) account, total power increased with uncut chip thickness, which can be seen from figure 5.

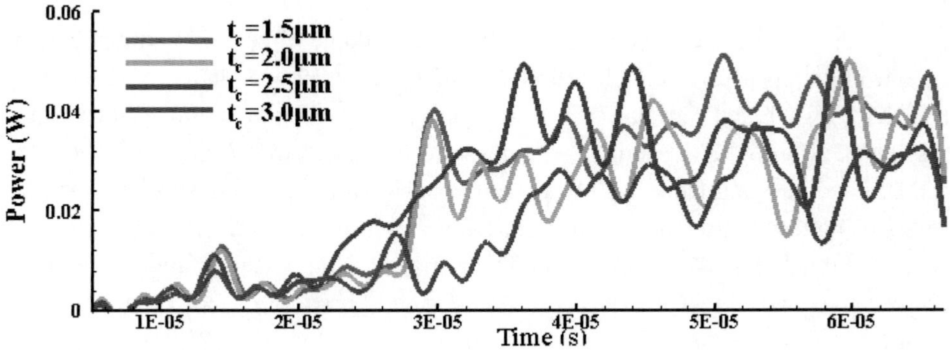

Figure 4. The total power for simulation

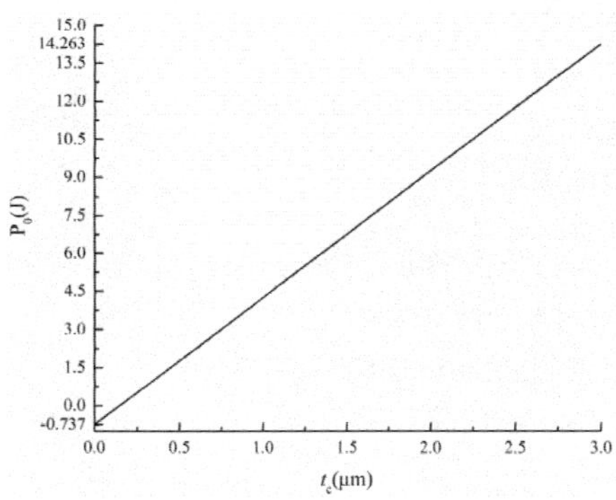

Figure 5. The total power for analysis results

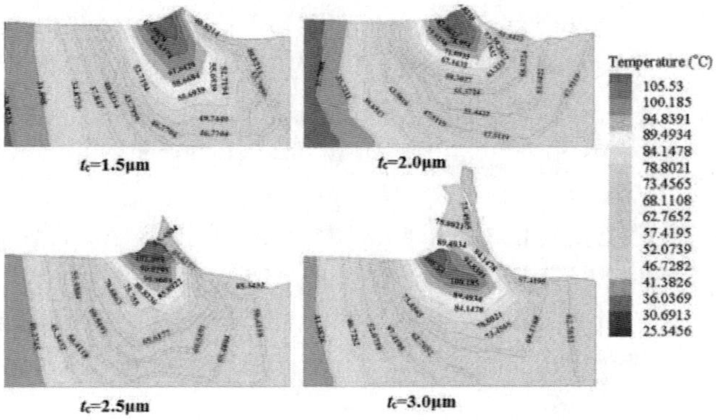

Figure 6. Workpiece temperature cloud at different uncut chip thickness

The workpiece's temperature cloud with different uncut chip thickness (t_c) as shown in figure 6. It can be noticed that chip formed was significantly influenced by the t_c. The larger t_c is, the lager chip is. The highest temperature is distributed in the primary deformation. And there is obvious temperature stratification phenomenon when close to the main deformation zone. But the most unfavorable point is the whole workpiece temperature is increasing, which will cause residual stresses and thermal deformation [11]. Therefore, to master the law of temperature changes and optimization of processing parameters is very necessary. According to this paper, although selecting bigger uncut chip thickness increased machining time, the workpiece will have a better improvement.

5. Conclusions

In this study, power model and 2-D FE simulations of micro milling have been developed. The developed model uses a semi analytical approach for calculating the power in the primary and secondary deformation zones. The results showed that the bigger uncut chip thickness is, the higger power and temperature are. Thermal-mechanical distribution of zone machined will reduced when the uncut chip thickness is larger than the minumum chip thickness.

Acknowledgments

This research is supported by the China Postdoctoral Science Foundation Funded Project (Grant No. 2016M592215), the Foundation for Outstanding Young Scientist in Shandong Province (Grant No. BS2014ZZ007), the Qingdao Postdoctoral Applied Research Project (Grant No.2015182) and the Scientific Research Foundation of the SDUST (Grant No. 2014RCJJ023).

Reference

[1] Yang K, Liang Y C and Zheng K N 2011 Tool edge effect on cutting temperature in micro-end-milling process *Int. J. Adv. Manuf. and Technol.* **52** 905-912

[2] Usui E, Kitagawa T and Shirakashi T 1978 Analytical prediction of three dimensional cutting process-Part 3: Cutting temperature and crater wear of carbide tool *J. Eng. Ind.* **100** 236–243

[3] Ulutan D, Lazoglu I and Dinc C 2009 Three-dimensional temperature predictions in machining processes using finite difference method *J. Mater. Process. Technol.* **209** 1111-1121

[4] Lazoglu I and Altintas Y 2002 Prediction of tool and chip temperature in continuous and interrupted machining *Int. J. Mach. Tools Manuf.* **42** 1011-1022

[5] Lazoglu I and Bugdayci B 2014 Thermal modeling of end milling *CIRP Ann. Manuf. Technol* **63** 113–116

[6] Lazoglu I and Islam C 2012 Modeling of 3D temperature fields for oblique machining *CIRP Ann. Manuf. Technol* **61** 127–130

[7] Thepsonthi T and Özel T 2015 3-D finite element process simulation of micro-end milling Ti-6Al-4V titanium alloy: Experimental validations on chip flow and tool wear *J. Mater. Process. Technol.* **221** 128-145

[8] Usui E, Hirota A and Masuko M 1978 Analytical Prediction of Three Dimensional Cutting Process-Part 1 Basic Cutting Model and Energy Approach *J. Eng. Ind.* **100** 222 -228

[9] Usui E and Hirota A 1978 Analytical Prediction of Three Dimensional Cutting Process-Part 2 Chip Formation and Cutting Force with Conventional Single-Point Tool *J. Eng. Ind.* **100** 229-235

[10] Molinari and Moufki A 2005 A new thermo-mechanical model of cutting applied to turning operations-Part I. Theory *Int. J. Mach. Tools Manuf.* **45** 166–180

[11] Zhang Q Y 2013 High-speed milling temperature and residual stress distribution study of 6061 aluminum alloy *Shanghai Jiaotong University*

Orthogonal Experimental Study on Tool Lifein Milling TB6 Titanium Alloy

LIU Dong [1,*], WANG Fang[1], WANG Jian min[1], XUE Yuan[1], XUE Jing[1]

College of Mechanical and Material Engineering, North China University of Technology, Beijing, 100144, China

liudong@ncut.edu.cn

Abstract. The tool wear was studied using orthogonal experiment method while milling titanium alloy TB6 in different milling parameters. The flank face wear was measured by the photomicrograph after a certain period of millingDuring the milling process. It can be seen from the differenceresults of the experiment that the most significant factor affecting the durability of titanium alloy TB6 cutter is the cutting speed, which difference is 130.67. Followed by the feed, the difference is 106.33; Next is the milling depth and its difference is 61.33; The least significant factor is the milling width.

1 Introduction

With the development of national defense industry, aerospace and other new technologies, the use and demand of titanium alloy materials are more and more extensive. Titanium alloy has excellent properties such as high specific strength, high specific modulus, lowexpansion coefficient, impact resistance, wear resistance, heat resistance and so on[1], therefore, titanium alloy has a good prospect. But its unique physical and chemical properties make titanium alloy a recognized refractory material[2], In the actual cutting process, the tool durability is low, the tool wear is serious, and the processing cost is high, which greatly hindered the application and promotion of this kind of materials[3]. The titanium alloy TB6 is a kind of titanium alloy near β type,which has high toughness, high strength and excellent machinability.The titanium alloy TB6 has higher specific strength and can greatly reduce the structural weight compared with the same strength structural steel or some other titanium alloys[4].

In the process of metal cutting, tool wear is a common concern[5]. Because it involves many aspects, it has a great influence on tool life, processing economy, machining accuracy, surface quality and reliability of machined workpiece. Many scholars at home and abroad have carried out a great deal of research on tool wear during cutting.Attanasio and other tools were used to predict the tool wear during the cutting of AISI1045 with carbide tools[6]. Filice and other tools were used to predict tool wear during the cutting of medium carbon steel[7]. ZhengMinliand other people, used uncoated carbide tool for titanium alloy cylindrical cutting experiment, founding the tool wear with adhesion wear, diffusion wear, and tool wear intensifiedwith the increase of cutting speed and feed rate[8]. ChenYan and other people used software on the tool in high speed cutting titanium alloy TC4 wear through finite element analysis and they found speeding up the tool wear with the increasing ofcutting speed, and when the cutting speed is 300m/min, tool life is only 1/3 of 130m/min[9].

In this paper, the titanium alloy TB6is subjected todry orthogonal cuttingunder different cutting parameters to process experimental research, comparative analysis of tool wear under different cutting parameters, andis subjected to regression analysiswith the experimental results to get regression formula of tool wear, and it provides some reference for the optimization of cutting parameters, improving tool life and reducing processing costs.

2 Experiment scheme

In this experiment, nine sets of milling parameters are selected according to the production experience, literature data and tool manufacturer's recommended value. Orthogonal cutting experiments are carried out, and the milling method is progressive milling. Experiment with nine identical tools. According to the recommended durability test standard, the flank wear amount VB=0.25-0.3mm is chosen as the tool wear standard. The wear of the tool is detected by a micro camera, machining the workpiece according to the experimental milling parameters, stoppingprocessing at regular intervals, the tool flank wear image is taken during the experiment, and processing wear images in software. Finally Current tool wear can be obtained.

The orthogonal experiment was adopted to design the experiment. Four factors and three levels of orthogonal experiments were designed. The four factors were milling speed, feed per tooth, cutting depth and cutting width. The level of each factor is shown in the following table 1.

Table.1 The cutting factor and level table

	level1	level2	level3
Cutting speed V_c m/min	15	25	35
Feed per tooth f_z mm/z	0.04	0.08	0.12
Depth of cut a_p /mm	1	1.5	2
Width of cut a_c /mm	0.12	0.17	0.22

3. Experimental condition

3.1 Experimental machine and Experimental tool

Fig.1 Experimental machine

Fig.2 Experimental tool

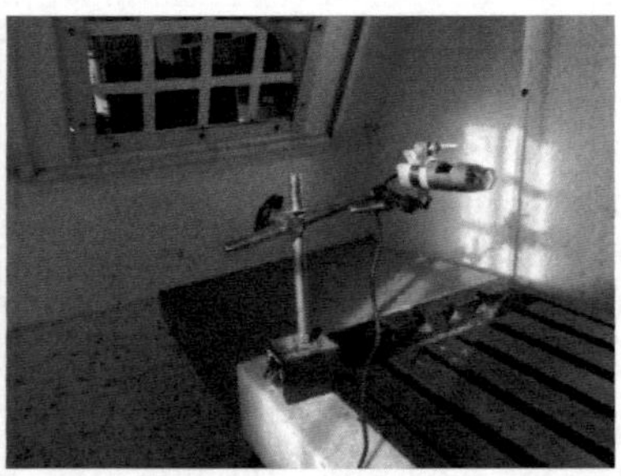

Fig.3 Experimental microscope

Fig.1 is machine tool used in experiments . Fig.2 is cemented carbide cutting tool used in experiments, 55° integral alloy end milling cutter, φ6R0.5*15*φ6*50L*55°*4T, Fig.3 was the experimental microscope camera.

3.2 Experimental material

TB6 cuboid material was used in the experiment. TB6 alloy, is also calledTi-1023. Ti-1023 is a typical titanium alloynearβtypewith a nominal component of Ti-10V-2Fe-3Al. The chemical composition and mechanical properties are shown in table 2 and table 3 below.

Table.2 Chemical composition of TB6 titanium alloy

Element	H	O	N	Fe	Al	V	Ti
Content（%）	0.01	0.03	0.03	1.93	2.95	10.16	other

Table.3 Mechanical properties of TB6 titanium alloy

Tensile strengthσ_b/MPa	Yield strength$\sigma_{0.2}$/MPa	Elongation rateδ/%	Shrinkage rateψ/%	Impact toughnessα_k/(J•cm^{-2})
1005	1035	6	64	60

3.3 Experimental instrument

Measure the wear of the tool by using a micro camera as shown as the Fig.3.The camera is connected with the computer, and the tool wear can be enlarged by using relevant software so as to be easily observed on the computer. Fix the camera on the machine tool and move the fitted tool to the right place to record the current coordinates. After a certain period of time, the cutting is stopped and the tool is moved to thebeforehandrecorded coordinates. The camera is used to observe and record the wear. If the desired amount of wear is not reached, the cutting can be continued for a certain period of time, then the cutting is stopped and the tool is moved to the beforehand recorded coordinate position. The wear condition is observed until the tool reaches a predetermined amount of wear.

4. Experimental results

According to the specified experimental scheme, the cutting time of each group is calculated as the wear criterion of the tool under the cutting parameters of each group, and the cutting time VB = 0.26mm is used to calculate the cutting time required for each group. As shown in the following table 4:

Table.4 Experimental results of tool durability

Serialnumber	Horizontal combination	Millingdepth $a_p$mm	Millingwidth $a_e$mm	Feed rate $f_z$mm/z	Cutting speed V_c m/min	Durability T min
1	A1B1C1D1	1	0.12	0.04	15	218
2	A1B2C2D2	1.5	0.17	0.08	15	190
3	A1B3C3D3	2	0.22	0.12	15	65
4	A2B1C2D3	1.5	0.22	0.04	25	144
5	A2B2C2D1	2	0.12	0.08	25	38
6	A2B3C1D2	1	0.17	0.12	25	25
7	A3B1C3D2	2	0.17	0.04	35	55
8	A3B2C1D3	1	0.22	0.08	35	18
9	3B3C2D1	1.5	0.12	0.12	35	8

The test specimen is a rectangular cubic titanium alloy, cutting in one direction after repeated cutting, record the number of times each group of repeated cutting and time, when the tool flank wear 0.26mm to stop cutting, at this time cutting Time is tool durability.

Figure 4 for a certain photomicrograph by the computer on the cutting tool when the wear and tear measurement, the figure at the tip of the white band with a tool wear, measured by the computer width.The difference is that the larger the difference is, the greater the influence of the parameter on the measured effect. The results of the above orthogonal experiments were used to calculate the average effect. The results of the analysis are shown in the table below.

The difference is that the larger the difference is, the greater the influence of the parameter on the measured effect. The results of the above orthogonal experiments were used to calculate the average effect. The results of the analysis are shown in the table 5 below.

The difference is that the larger the difference is, the greater the influence of the parameter on the measured effect. The results of the above orthogonal experiments were used to calculate the average effect. The results of the analysis are shown in the table 5 below.

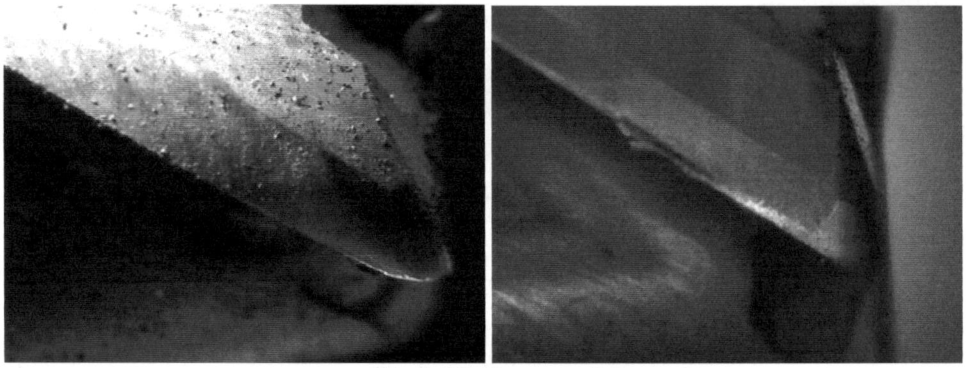

Fig.4 Flank wear of the tool

Table 5. The impact results VS cutting parameters

	t			
	a_p	a_e	f_z	V_c
K_1	87	88	139	157.67
K_2	114	90	82	69
K_3	52.67	77.67	32.67	27
R	61.33	12.33	106.33	130.67

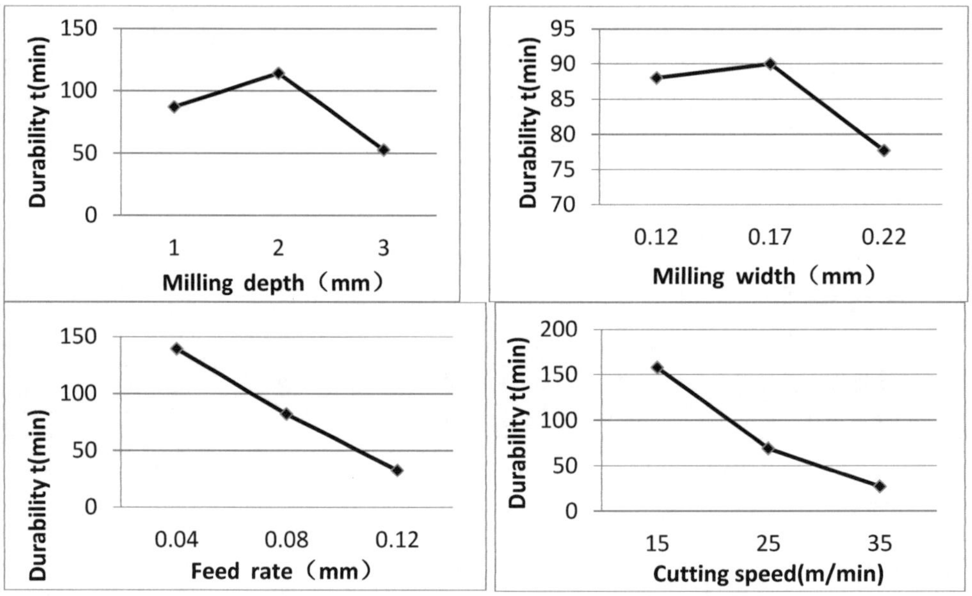

Fig5 Durability vs. change of cutting parameters

From the experimental results can be seen in the titanium alloy TB6 knife milling, the most significant impact on the durability of the cutting speed, the difference is 130.67; followed by the feed, the difference is 106.33; followed by Milling depth, the difference is 61.33; the most significant factor for the milling width, the difference is 12.33. The impact of factors on the durability of the impact of the effect of Fig5 shown.

It can be seen from the Fig 5 that the durability decreases with the increase of feed rate and cutting speed. The reason is that the higher the milling speed is, the higher the friction, the higher the milling temperature, the faster wear and the durability. The increase in feed volume to make the chip wide, conducive to heat conduction, speed up the temperature increase, thereby reducing the durability.

Durability increases first and then decreases with increasing milling depth. The durability increases first and then decreases as the milling width increases.

5 Conclusion

(1) By the above experimental results and analysis can be concluded that, when using TB6 titanium alloy cutting tools, milling dosage effect on durability according to a significant degree of milling speed, feed, milling depth, milling width; The durability decreases with the increase of feed quantity and cutting speed, and the durability decreases with the increase of the depth of milling. Durability decreases as the width of milling increases first.

(2) Milling depth and milling width have less impact on durability than milling and feed. So with titanium alloy TB6 tool, the use of small feed, small milling depth, low milling speed can be very high durability, improve tool life, take a good cutting effect.

Acknowledgments

Authors wishing to acknowledge assistance of the North China University of Technology university student science and technology activity project.

References

[1] HU Hong-nan , DONG Ming. New Development on Cutting Technology of Particles Enforcement Metal Matrix Composites[J]. Metal materials and Metallurgical Engineering, 2009, 37(1):3-7.
[2] Chen Wu-yi, Yuan Yue-feng. Research Development of Cutting Technology for Titanium Alloy[J]. Aeronautical Manufacturing Technology, 2010(15):25-30.
[3] ARAMESH M,SHI B,NASSEF A O, et al.Meta-modeling optimization of the cutting process during turning titanium metal matrix composites[J].ProcediaCIRP,2013,8:576-58
[4] WANG Mei-jiao,MENG Xiang-jun,LIAOZhi-qian. Research status of Ti-1023 alloy [J]. Material development and Application, 2009,24(5):66-69.
[5] ZHANG Zong-yang,AIXing,LIU Zhan-qiang. Study on wear characteristics of high-efficiency cuttingiron-based superalloy [J].Manufacturingautomation,2011,10:48-52.
[6] ATTANASIO A,CERETTI E,FIORENTINO A, et al.Investigation and FEM-based simulation of tool wear in turning operations with uncoated carbide tools[J].Wear,2010,269(5-6):344-350.
[7] FILICE L,MICARI F,SETTINERI L, et al. Wear modeling in mild steel orthogonal cutting when using uncoated carbide tools[J]. Wear,2007,262(5-6):545-554.
[8] ZHENG Min-li,FAN Yi-hang. An Overview of Tool Friction and Wear Behavior in High-speedMachining-typical Difficult-to-cut Material[J]. Journal of Harbin Institute of Technology,2011(6):22-30.
[9] CHEN Yan, YANG Shubao, FU Yucan. FEM estimation of tool wear in high speed cutting of Ti6A14V alloy[J]. Journal of Aeronautics, 2013,34(9):2230-2240.

IOP Conf. Series: Materials Science and Engineering **265** (2017) 012019 doi:10.1088/1757-899X/265/1/012019

Numerical simulation of hull curved plate forming by electromagnetic force assisted line heating

Wang Ji[1,2], Wang Shun[1]*, Liu Yujun[1], Li Rui[1], Liu xiao[1]
[1]School of Naval Architecture, Dalian University of Technology, Dalian, China
[2]Collaborative Innovation Center for Advanced Ship and Deep-Sea Exploration, Shanghai, China

*Corresponding author: wangshun@mail.dlut.edu.cn

Abstract. Line heating is a common method in shipyards for forming of hull curved plate. The aluminum alloy plate is widely used in shipbuilding. To solve the problem of thick aluminum alloy plate forming with complex curved surface, a new technology named electromagnetic force assisted line heating(EFALH) was proposed in this paper. The FEM model of EFALH was established and the effect of electromagnetic force assisted forming was verified by self development equipment. Firstly, the solving idea of numerical simulation for EFALH was illustrated. Then, the coupled numerical simulation model of multi physical fields were established. Lastly, the reliability of the numerical simulation model was verified by comparing the experimental data. This paper lays a foundation for solving the forming problems of thick aluminum alloy curved plate in shipbuilding.

1. Introduction

Line heating is the main technique of manufacturing large hulls and complex curved plate, which has the advantages of wide ranges of application, low cost, flexible and suitable for a variety of complex curved shapes. In advanced shipbuilding countries such as Japan, Korea and so on, this technique is supported by many funds and investments to research on the deformation mechanisms of curved plate with different heat sources, deformation rules, prediction method of heating parameters and automatic machining equipment[1, 2]. However, there are lots of influence factors in traditional line heating process which results in the difficulties of getting the forming regularity and controlling the process accuracy[3]. In addition, high heating temperature may change the plate material even generate cracks. The weight control of hull is very important in shipbuilding. Comparing with the traditional steel plate, the aluminum alloy has advantages including lower density, higher strength, better corrosion resistance and better welding property. The aluminum alloy plate is widely used in shipbuilding.

The current metal plate forming methods can be divided into cold forming methods and hot forming methods. Cold forming method including stamping forming[4], electromagnetic forming[5] and laser shock peening method[6]. In stamping forming, the material is extruded into specific shapes by stamping machine. However, this method can be only applied to thin plate and is not suitable for thick plate for ship manufacturing. The electromagnetic forming is wildly used in small component machining in vehicle industry. The molds of this method are very expensive. The shapes of the components in ship hulls are varied. It is not suitable for mass production, which makes the cost of application of electromagnetic forming in ship industry is really high. For the laser shock peening method, the deformation of each peening is very small. However, if the peek pressure is increased to get a bigger deformation, the finishing surface will be very rough and the property of the material is influenced[7].

The hot forming methods include plate warm forming technology and line heating[8]. The plate warm forming technology is mostly applied in the vehicle and aeronautics engineering. It cannot be applied in large components manufacturing in ship industry[9]. Guo, et.al[10] studied the line heat process for aluminum alloy and found that the suitable processing temperature for aluminum alloy is between 200℃ and 300℃. However, under this temperature, it is very hard to get enough deformation for thick aluminum plate. To overcome these challenges, people put forward some new line heating forming technologies such as induction heating forming[11], laser heating forming[12]. However, there is no new line heating technology can well meet the requirements of aluminum alloy plate until to now. Induction heating is prone to overheat on the edge of the plate and especially has low energy utilization. Laser heating need large and expensive equipment.

Both cold and hot traditional processing methods are not suitable for aluminum alloy plate forming of curved plate. This paper provides an advanced method which combines the electromagnetic forming process and line heating, which is named electromagnetic force assisted line heating(**EFALH**). It means that during the line heating process, the temperature is controlled in a suitable range while electromagnetic force is applied to the aluminum alloy plate to get enough deformation. In this paper, the coupled numerical simulation model of multi physics fields are established. The experiment is conducted to measure the data including the wave form of the pulse, the local deformation of the plate. The reliability of the numerical simulation model can be verified by comparing the experimental data with the numerical simulation results.

2. The solving idea of numerical simulation

In traditional line heating forming process, moving flame and cooling water are needed, as shown in Figure 1(a). The most common kinds of heating source in shipyards are acetylene and propylene. The forming theory of line heating forming was well introduced in previous researches. So this paper will only highlight on the introduction of EFALH. In this method, an external force is added to the plate forming process by line heating, as shown in Figure 1(b). The electromagnetic coil is placed above the steel plate and the coil is kept in the appropriate position to the heat source during processing. When the pulsed current flows through the electromagnetic coil, a mutative magnetic field can be generated. The magnetic field will generate a vortex which is forced by the magnetic field. Then the radial electromagnetic force that acts on the heating line by extrusion and the axial magnetic pressure that promotes angle deformation of plate are generated at the same time. This device can control the deformation of the plates by relying on the evolution of the stress and strain.

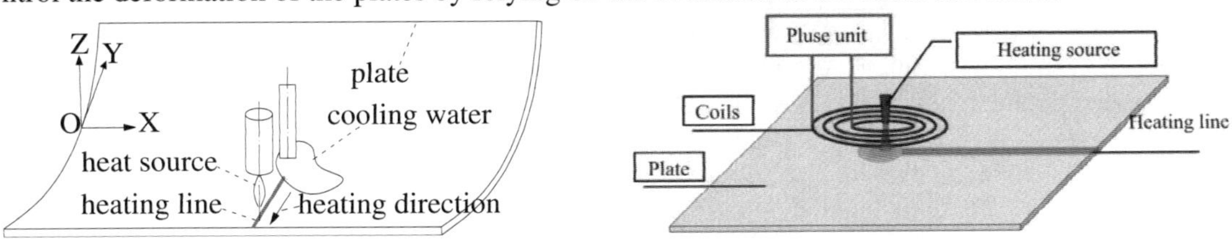

(a) line heating process (b) EFALH process

Figure 1 Process of line heating and EFALH

The simulation model includes the circuitry, the electromagnetic field, the temperature field and the structure field. So a suitable coupling method is very important and will determine the accuracy of the simulation result. The relationship between each fields is not linear, so finite element analysis are used to achieve the multi physical field coupling. The circuitry is used to provide the pulse current to the electromagnetic coils. The process can be achieved by circuitry- electromagnetic coupling. On the other hand, the resistivity of the plate changes with the temperature. So the influence of the temperature to the electromagnetic force is significant. The deformation of the plate can be influenced by the heat source, the electromagnetic force and structure. However, the small deformation cannot influence the temperature and electromagnetic field too much, so only thermo-structure and electromagnetic-structure one-way coupling are analyzed.

Figure 2 shows the research route of the simulation model. At first, all the elements including the aluminum plate, the line heat resource are the electromagnetic coils are built in the Three-dimensional model. The electromagnetic field, the temperature field and the structure field can be calculated. With the temperature field and the resistivity curve of the aluminum material and the electromagnetic field, the thermo-electromagnetic coupling analysis can be achieved and the electromagnetic force can be calculated. With the structure field, the final deformation of the aluminum plate can be calculated.

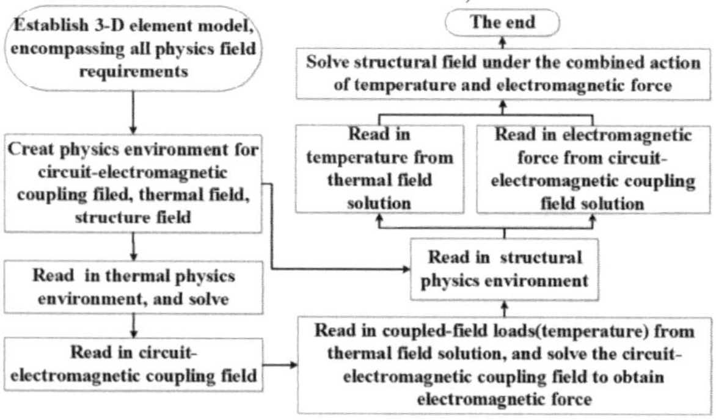

Figure 2. Flow chart of multi-physics couplings of EFALH

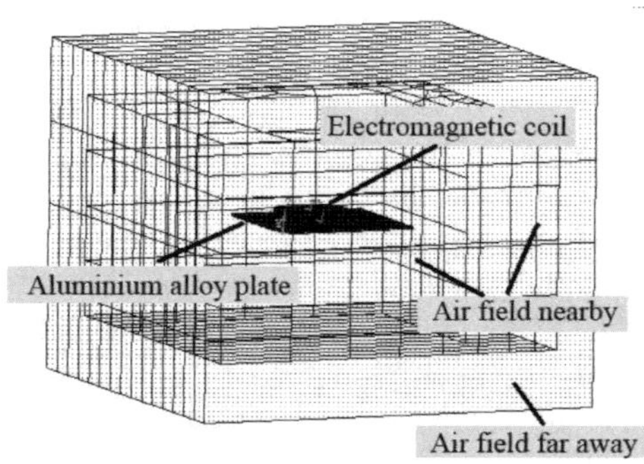

Figure 3. FEM model of EFALH

3. Three-dimensional finite element model of EFALH

The plate material is aluminum alloy. The temperature fields, the electromagnetic fields and the structure fields are effected by heating time and heating position. In the model, the deformation, the force and the temperature of different location and time should be simulated under the effect of the multi physical fields coupling. Two-dimensional finite e element model cannot satisfy the requirement, so three-dimensional FEM are used in the simulation. The FEM model of EFALH including the whole structure is shown in Figure 3.

In the simulation model of the circuitry-electromagnetic coupling, the circuitry part includes the individual current I0, the resistance R0 and the coils P0. Because the current is equal at different location in the coils, two same circuitries are used to couple the two layers of the coils. The current I0 is defined as pulse current and is provided to the coils.

4. Verification of the EFALH numerical model

The propose of the verification experiment is to test the effect of EFALH and accuracy of EFALH numerical model. The plate material is aluminum alloy 5083 H116. The plate length and breadth is 2000mm $\times$ 1000mm. The plate thickness is 8mm and 10mm, heating time is 55 seconds and 80 seconds, respectively. The heating length is 300mm. The flow rate of propylene and oxygen is determined according to forming experience[8]. The location of heating line is at the side of plate. The experimental equipment of electromagnetic force assisted line heating is shown in Figure 4. The shrinkage was measured by digital caliper. The measuring method refer to reference[8]. The local deformation is instead by the form of local shrinkage. The pulse current is measured during the EFALH process. The data of local shrinkage for EFALH and line heating process are measured and recorded before and after forming.

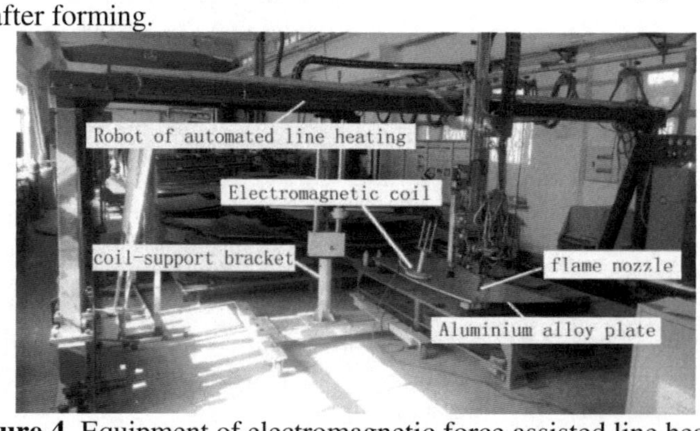

Figure 4. Equipment of electromagnetic force assisted line heating

4.1 Verification of pluse current

The pulse current in the electromagnetic coil of experiment is measured by rogowski coil and oscilloscope. The numerical and measuring results of pulse current in the electromagnetic coil are shown in Figure 5. The current rises and drops rapidly in both the numerical and measured results. It can verify the accuracy of the circuitry-electromagnetic coupling model.

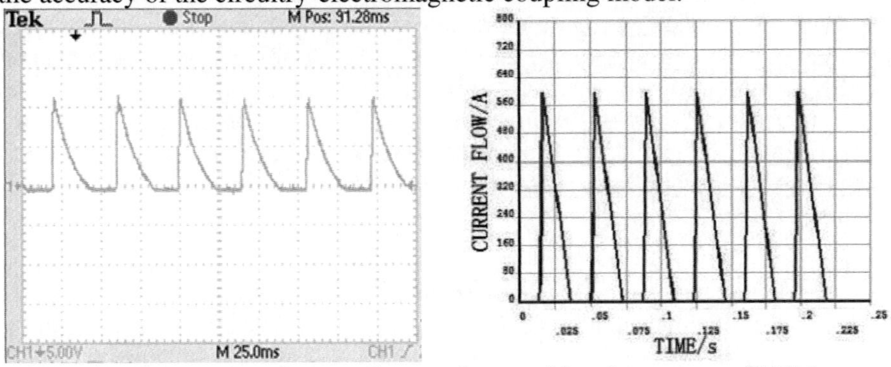

(a) pulse current of measuring result (b) pulse current of FEM

Figure 5. Numerical and measuring results of pulse current in the electromagnetic coil

4.2 Verification of deformation

The parameters of the pulse current are shown in Table 1. The distance between the aluminum plate and the electromagnetic coils is 5mm. The input data in the simulation is according to the experiment parameters under different conditions. The heating time is 55seconds for 8mm thickness plate and 80 seconds for 10mm thickness plate.

The FEM and experiment results are shown in Figure 6. The deformation of EFALH process is bigger than the single line heating process of the same experiment condition. The electromagnetic force increases the shrinkage of the curved plate. It shows that the EFALH process has a better forming performance than the single line heating process. On the other hand, the differences between the numerical and experiment results under both experiment conditions are very small, in which the accuracy of the EFALH numerical model is proved.

Table1. The parameters of the pulse current

No.	Parameters	Value
1	Pulse current amplitude	750A
2	Pulse current frequency	10.86Hz
3	Rising time of the pulse current	1.2ms
4	Dropping time of the pulse current	34ms
5	Peak pulse duration	0.001ms

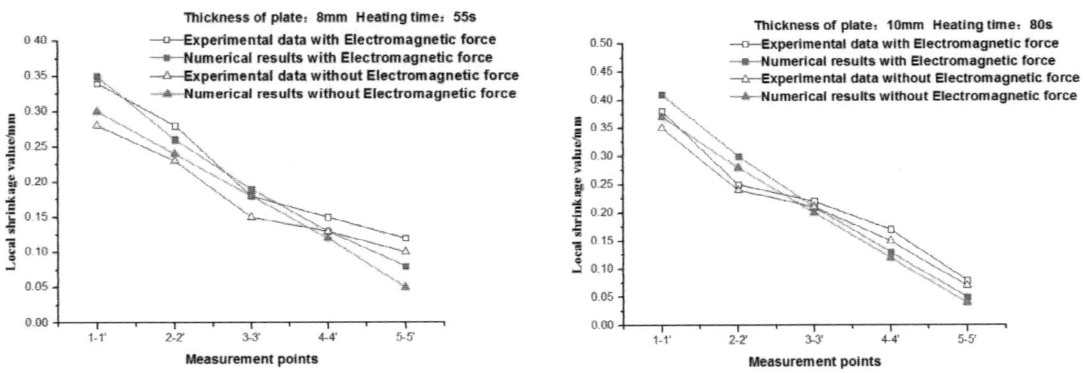

（a）plate with 8mm thickness （b）plate with 10mm thickness

Figure 6. Local shrinkage of EFALH and line heating process

IOP Conf. Series: Materials Science and Engineering **265** (2017) 012019 doi:10.1088/1757-899X/265/1/012019

5. Conclusion

A new technology named EFALH was proposed in this paper to overcome the forming difficulties of thick aluminum alloy plate. The FEM model of EFALH was established and the effect of electromagnetic force assisted forming was verified by means of self-development equipment. The electromagnetic force assisted line heating process can increase the efficiency of traditional method while it can also keep the temperature in a suitable range. The finite element analysis model with multi physical fields coupling including circuitry-electromagnetic coupling, thermo-electromagnetic one way coupling, thermo-structure and electromagnetic-structure one way coupling was established. Experiment was conducted to verify the accuracy of the FEM model of EFALH. Both the simulation and the experiment results shows that EFALH has a better forming performance than traditional line heating method. This paper lays a basis for solving the forming problems of thick aluminum alloy curved plate in shipbuilding.

Acknowledgment

This study was supported by the National Natural Science Foundation of China(No. 51009025) and National Science and Technology Major Project of the Ministry of Science and Technology of China(2013ZX04003041).

References

[1] Biswas P., Mandal N., Sha O. Numerical and dimensional analysis for prediction of line heating residual deformations. Journal of Marine Science and Application[J]. 2010,9(1): 14.

[2] Ryu C., Shin J. G. Optimal Approximated Unfolding of General Curved Shell Plates Based on Deformation Theory. Journal of Manufacturing Science and Engineering[J]. 2006,128(1): 261.

[3] Wang S., Liu Y., Wang J., et al. Forming laws and model for line heating considering whole deflection. Huazhong Keji Daxue Xuebao (Ziran Kexue Ban)/Journal of Huazhong University of Science and Technology (Natural Science Edition)[J]. 2016,44(7): 30-35.

[4] Wu H. Simulation and experimental study of varibale thickness blanks stamping forming[D]. Shanghai: Shanghai Jiao Tong University, 2014.

[5] Oliveira D. A., Worswick M. J., Finn M., et al. Electromagnetic forming of aluminum alloy sheet: free-form and cavity fill experiments and model. Journal of Materials Processing Technology[J]. 2005,170(1): 350-362.

[6] Ivetic G. Three-dimensional FEM analysis of laser shock peening of aluminium alloy 2024-T351 thin sheets. Surface Engineering[J]. 2011,27(6): 445-453.

[7] Bagheriasl R. Formability of aluminum alloy sheet at elevated temperature. [J]. 2012.

[8] Wang J. Research on pivotal techniques for automatic line-heating process[D]. Dalian: Dalian University of Technology, 2007.

[9] Xu X. F. Research on mechanical behavior and simulation and experiment of superplastic forming of 5083 aluminum alloyNanjing University of Aeronautics and Astronautics.

[10] Guo X., Jing B., Yan Y., et al. Technics of Fire and Water Rectification for Thick Plate of Aluminium Alloy. Development and Application of Materials[J]. 2013,(01): 26-28.

[11] Lee K. S. Deformation Behavior of Doubly Curved SS400 Thick Metal Plates by High Frequency Induction-Based Line Heating. Key Engineering Materials[J]. 2013,535-536: 369-372.

[12] Siqueira R. H. M., Carvalho S. M., Kam I. K. L., et al. Non-contact sheet forming using lasers applied to a high strength aluminum alloy. Journal of Materials Research and Technology[J]. 2016.

IOP Conf. Series: Materials Science and Engineering **265** (2017) 012020 doi:10.1088/1757-899X/265/1/012020

A method for increase abrasive wear resistance parts by obtaining on methods casting on gasifying models

V V Sedukhin[1], A N Anikeev[1] and I V Chumanov[1]

[1] Department of engineering and technology of materials, South Ural State University (national research university), 16 Turgenev Street, Zlatoust, Russia

Abstract. Method optimizes hardening working layer parts', working in high-abrasive conditions looks in this work: bland refractory particles WC and TiC in respect of 70/30 wt. % prepared by beforehand is applied on polystyrene model in casting' mould. After metal poured in mould, withstand for crystallization, and then a study is carried out. Study macro- and microstructure received samples allows to say that thickness and structure received hardened layer depends on duration interactions blend harder carbides and liquid metal. Different character interactions various dispersed particles and matrix metal observed under the same conditions. Tests abrasive wear resistance received materials of method calculating residual masses was conducted in laboratory' conditions. Results research wear resistance showed about that method obtaining harder coating of blend carbide tungsten and carbide titanium by means of drawing on surface foam polystyrene model before moulding, allows receive details with surface has wear resistance in 2.5 times higher, than details of analogy steel uncoated. Wherein energy costs necessary for transformation units mass' substances in powder at obtained harder layer in 2.06 times higher, than materials uncoated.

1. Introduction

Open pit is considered most effective way pray useful fossils. Process excavation appear most important element technological chain open pit, which continuity largely determined durability exchangeable teeth buckets quarry excavators, details drilling equipment etc., which experience direct interaction with the rock. Details working in extreme abrasive wear conditions during excavation and exploitation particularly strong and abrasive solids (a large number of favourably oriented cutting edges on the surface fragments blasted rock, the dominant role metal micro cutting, intense impact loads, etc. [1]), which causes their rapid failure [2,3].

Analyses failure details different appointments showed, that, despite on different conditions their works, the main reason failure serve wear into abrasion process in interaction with abrasive material [4]. Hardening by dispersed parts introducing there in crystallizing liquid and deposition hardened layer a certain thickness on the surface details can be method increase mechanical characteristic materials [5-7]. The letter method has been used as a basis in this work.

2. Execution of works to obtain experimental' samples

Method was offered for increase wear resistance working edges, consisting in that mechanical hardening blend applying on the surface foam polystyrene model, which is further working. The proposed coating represents blend fine powders (2-5 μm) carbide tungsten and carbide titanium. It provides simplicity and cheapness production blend. The ratio of components is (in wt. %) - 68:28:4 (WC:TiC:parafine).

An experiment was been on samples of arbitrary shape as approbation of the method. Details received of methods lost foam casting. This technology implies covering polystyrene models' anti-weld blends, which is protecting the metal from fumes to moulding sand. Working part details' before molding was covered mechanical strengthening blend. A canal was cutter out into polystyrene models for ensuring the better interaction liquid metals and coating. Gating system was covered anti-weld paint', after the bland has solidified, then dried. Received models were formed in sand before dried. Steel AISI 1020 used as basic metal. Choice of steel was due to minimal content' carbon and alloying components, which can influence on the process at the interaction of the strengthening mixture and basis metal details'. 3 casting obtained during the experiment ('figure 1').

Figure 1. Typical configuration samples obtained on results experiment.

3. Study of the structure of samples

Study of the structure obtained samples with the help of an inverted metallographic microscope 'C. Zeizz Observer. D1m' equipped with an image analyzer 'Thixomet.PRO' [8], as well as using an electronic scanning microscope 'JEOL JSM–6460 LV'. Structure study revealed, that dispersed particles titanium carbides' and tungsten in blend, reacted with the liquid metal and formed a coating 'figure 2'. The interaction of these components with the liquid melt occurred, but complete dissolution of tungsten carbides (lights) and carbides titanium (dark) not observed 'figure 3, 4'. It's related with high speed crystallization metal. Chemical composition in volume samples represented in table 1.

The complete dissolution of hardening particles in the surface layer castings 'figure 5', and not their finding in the form separate phases, is observed when studying the microstructure of the samples obtained in the course repeated experiment. The remaining carbide particles, which did not have time to dissolve in the volume of the metal, are observed at large magnifications 'figure 6'. The presence non-metallic inclusions in it, containing a significant amount (20-22 wt. %) of titanium also indicates the dissolution particles in the surface layer indicate 'figure 7'. The chemical composition detected particles in the volume of the samples obtained during the second experiment is shown in table 2. The inconsistent compositions and behaviour of the dispersed particles of tungsten carbide and titanium carbide on interaction with the metallic melt in the first and second experiments are probably related to their different wettability characteristics: under the same conditions, the nature and phenomenon of the given process for different particles differs, as shown by the authors of [9-13].

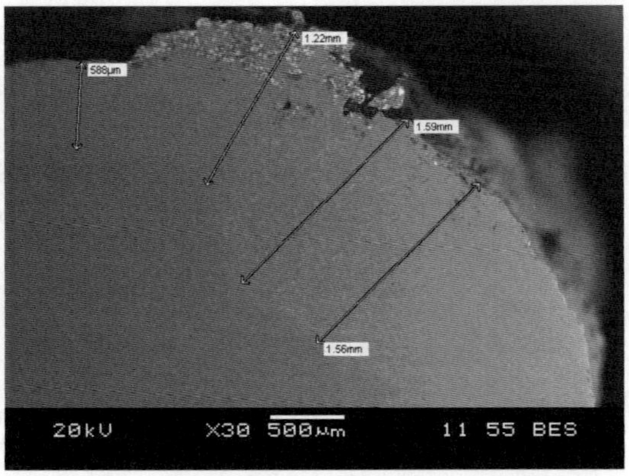

Figure 2. Outer layer samples, obtained during the experiment, ×30.

Figure 3. Detected particles of tungsten carbides in samples, obtained during the experiment, ×500.

Figure 4. Detected particles of titanium carbides in samples, obtained during the experiment, ×500.

Table 1. Chemical composition detected particles.

	Content of elements (wt. %)			
	C	Ti	Fe	W
3	10.98	-	-	89.02
4	14.62	84.40	0.98	-

Figure 5. The outer layer samples, obtained during the repeated experiment, ×100.

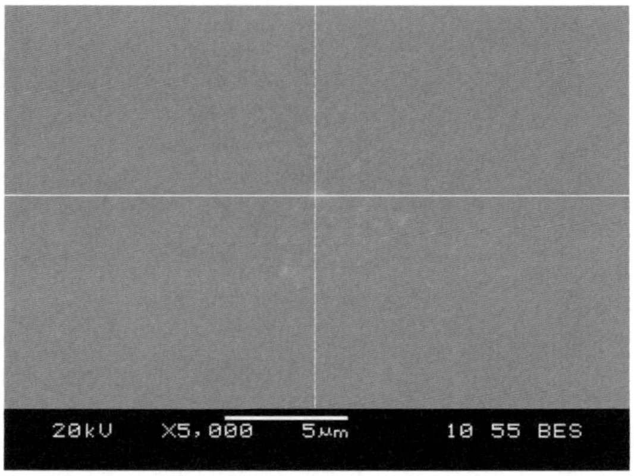

Figure 6. Detected particles of tungsten carbides in samples, obtained during the repeated experiment, ×5000.

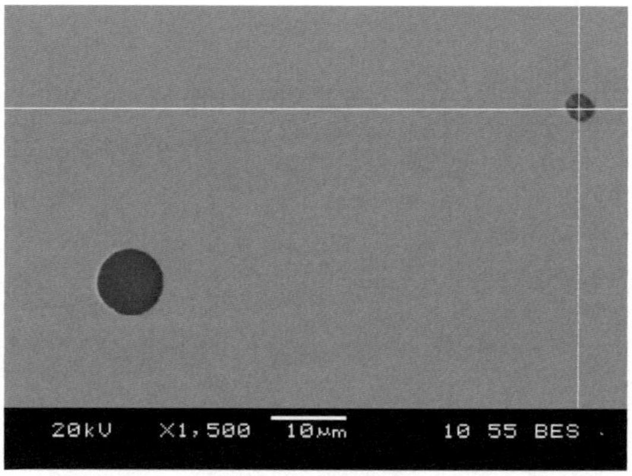

Figure 7. Detected inclusions with titanium content in samples, obtained during the repeated experiment, ×1500.

Table 2. Chemical composition detected particles.

	Content of elements (wt. %)								
	C	O	Si	P	S	Ti	Cr	Fe	W
6	7.01	-	-	6.31	-	-	3.69	63.31	19.68
7	3.51	17.94	8.11	-	5.53	21.67	-	43.24	-

4. Study of wear resistance samples

The wear resistance of the samples was investigated under laboratory conditions by weighing the residual masses. The study of abrasion resistance was carried out by abrasive abrasion with a certain length of abrasive tape on which the sample was placed under load; the study was conducted in 2 iterations. Wear resistance was measured on the samples obtained during the repeated experiment, with the surface of the obtained sample, the inverse of the "working" (on which the reinforced layer was obtained), as the standard. Also, after measuring the weight loss of the test samples, the energy spent on converting a unit of mass of substances into a powder (A, [MJ/kg]) was calculated.

Table 3. Results of measurement of wear resistance.

	Reference	The resulting sample
m_0, g	4.6864	4.6520
m_1, g	4.6845	4.6512
m_2, g	4.6826	4.6505
Average weight loss	0.002	0.0008
A, MJ/kg	4.1	8.45

5. Conclusion

According to the obtained data, can be observed that the method for obtaining hardened coatings from a bland of tungsten carbide and titanium carbide by applying it to the surface of a polystyrene model before moulding into casting moulds allows producing parts with a surface having a wear resistance 2.5 times higher than that of parts made from Similar to the unalloyed steel grade. At the same time, the energy expenditure necessary to convert a unit of mass of substances into powder from the resulting hardened layer is 2.06 times greater than for a similar material without coating. At the same time, the method of obtaining a reinforcing mixture is simple and cheap. The main criterion, which must be paid attention in the production of parts - a sufficient supply of liquid melt to the details. This is necessary for a longer crystallization time of the metal and, correspondingly, for a longer time of its interaction with dispersed refractory particles.

Acknowledgments

South Ural State University is grateful for financial support of the Ministry of Education and Science of the Russian Federation (grant No 11.9658.2017/8.9.)

References

[1] Sokolova N M and Belova M E 2008 *Factory laboratory. Diagnosis of materials* **74** 61
[2] Dragobetskii V V, Shapoval A A, Mospan D V, Trotsko O V and Lotous V V 2015 *Metallurgical and Mining Industry* **7** 363
[3] Dobrovolskiy A G and Koshelenko P I 1989 *Abrasive wear resistance of materials* (Kiev: Technics)
[4] Mulyavko N M 2001 *Proceedings of the Chelyabinsk Scientific Center* **4** 28
[5] Chumanov I V, Kareva N T, Chumanov V I and Anikeev A N 2012 *Russian Metallurgy (Metally)* **2012** 539
[6] Chumanov I V, Chumanov V I and Anikeev A N 2015 *Indian Journal of Science and Technology* **8** 7
[7] Anikeev A N, Seduhin V V and Sergeev D V 2016 *Materials Science Forum* **843** 269
[8] Internet-resource: http://www.thixomet.ru/products/?show=2 (date of the application 29.07.2017)
[9] Passerone A, Muolo M L and Valenza F 2016 *Journal of Materials Engineering and Performance* **25** 3330
[10] Xi L, Kaban I, Nowak R, Bruzda G, Sobczak N, Stoica M and Eckert J 2016 *Journal of Materials Engineering and Performance* **25** 3204
[11] Sobczak N, Purgert R M, Asthana R, Sobczak J J, Homa M, Nowak R, Bruzda G, Siewiorek A and Pirowski Z 2016 *Ceramic Engineering and Science Proceedings* **36** 309
[12] Asthana R and Sobczak N 2014 *Ceramic Transactions* **248** 591
[13] Gambaro S, Valenza F, Passerone A, Cacciamani G and Muolo M L 2016 *Journal of the European Ceramic Society* **36** 4185

IOP Conf. Series: Materials Science and Engineering **265** (2017) 012021 doi:10.1088/1757-899X/265/1/012021

Effect of Co - based Alloy on Properties of Laser Cladding Layer

Y Yang, Z P Jiang, H Z Li

College of Mechanical Engineering, Qingdao University of Technology, Qingdao, China

Abstract. A large number of laser cladding experiments have been carried out using 20CrMnTi steel as substrate and Co-based alloy as cladding material. The influence of Co-based alloy on the laser cladding properties of 20CrMnTi steel was studied by analyzing the macroscopic and microscopic characteristics of cladding crack susceptibility, dilution rate, microstructure and friction and wear properties. The results show that the high-power laser cladding of Co-based material can obtain a flat defect-free cladding layer with compact structure and low crack susceptibility. A multi-layer cladding strategy with variable power can be used to fabricate thin wall structures without collapse Parts, the surface smooth without pores.

1. Introduction

Laser cladding technology is a new type of surface modification technology developed with the development of high-power lasers. It is the alloy powder is placed on the surface of the cladding substrate with different filler, the alloy powder is rapidly melted by laser as heat source, The surface of the substrate to form a coating with excellent performance, so as to achieve the purpose of strengthening the surface of the workpiece [1]. Laser cladding remanufacturing forming technology is in line with the national recycling economy and sustainable development strategy of green manufacturing technology, is also one of the state's key high-tech support [2-3]. 20CrMnTi is a kind of excellent carburizing steel with good hardenability, high impact toughness at low temperature and good fatigue resistance. It is widely used to manufacture various gears, shafts and piston parts. It is widely used in industry [4-6], to study the material remanufacturing process has a certain practical significance.

A lot of researches have been done on laser cladding technology. Valsecchi [7] and so on using the optimal process parameters and argon protection conditions were prepared under the conditions of WC content of up to 85% of the Co-based cladding layer, to achieve a high surface hardness and low crack sensitivity combination. Farnia et al [8] prepared Co-based Stellite 6 coatings on low carbon steel substrates and studied the growth mechanism of the grains. The growth of grains in the Co-based cladding material was studied by the experiments of starting austenite Tissue grains.

However, for 20CrMnTi steel materials, the laser cladding alloy powder research has not been reported. Which material is more suitable for 20CrMnTi steel parts laser cladding process, need further study and analysis. The influence of Co-based alloy on the laser cladding properties of 20CrMnTi steel was studied in this paper, which provided the basis for choosing suitable alloy powder for 20CrMnTi laser cladding process.

2. Laser cladding test

The carburized 20CrMnTi steel was used as the substrate, and the size was Φ170 mm × 10 mm. In order to ensure the consistency of the test and the actual parts of the state, the matrix sample carburizing quenching heat treatment, carburizing layer of effective hardened layer depth of 0.8 ~ 1.2

mm. Before the test, the substrate sample surface sanding to remove the surface oxide; use acetone to clean the residual stains after drying.
The main components and content (mass fraction,%) are: C 1.15, Mn 0.50, Mo 0.80 ~ 1.10, W 4.00, Fe 3.00, Co, Si 1.10, Ni 3.00, Cr 29, Co balance. The alloy powder was stored in a vacuum atmosphere before the test.
Laser DF4000-60 semiconductor laser was used in the experiment. Laser spot diameter was 4 mm. Powder feeder model SF02 carrier gas synchronous powder feeder, powder feeding gas and high purity nitrogen gas protection. The laser parameters were as follows: laser power 2800w, scanning speed 180, 360 and 540 mm/min; laser power 3500W, scan speed respectively, the parameters were as follows: single-channel cladding test, preset 2 mm alloy powder, 180, 360 and 540 mm / min. The laser power was 4200 w and the scanning speed was 180, 360 and 540 mm / min, respectively. The laser parameters were: laser power 3500 W, scanning speed 180 mm / min, powder feeder voltage 9 V, overlap rate of 33%, 50 m / min. % And 67% respectively. 1 set of multi-layer stacking cladding test, test parameters 1 to 10 layers, laser power 1500 w; 11 to 15 layers, laser power 800 w.
The surface quality, macroscopic crack and cross-sectional morphology of the cladding layer were observed by DTX stereomicroscope. The microstructure of the cladding layer was observed by using electronic metallographic microscope. The etchant was Nitro-hydrochloric acid. The hardness of the cladding layer was measured by FM-700 micro-hardness tester.

3. Test results and analysis

3.1. Cladding crack susceptibility

Crater at the topography shown in Figure 1. The research shows that the crater is jagged and the speed of arc welding at the end of the welding process is quicker to cool the molten pool and the metal can be crystallized in a very short time. But the liquid metal can not meet the need of final crystallization. , Leading to cracks, so the crater cracks are hot cracks. The physical process of laser cladding and welding have certain similarities, all of which are heated and melted to form molten pool and then solidified. Therefore, the formation mechanism of the crack can be referred to the conclusion of welding process. From Figure 1, Which is in accordance with the character of crack in welding. The reason is the end stage of cladding, the laser beam disappears, the pool rapidly cools and the metal rapidly solidifies to form crater crack.

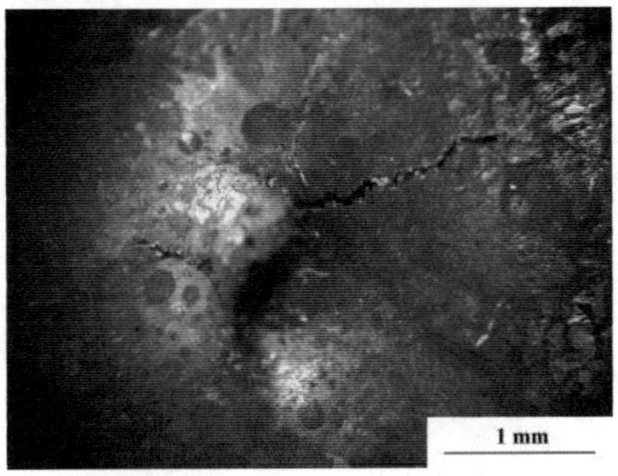

Figure 1. Crater morphology

3.2. The dilution rate of the cladding layer

The cladding layer and the matrix material to form a good metallurgical bonding is to ensure that the cladding layer in use without cracking, shedding of the important conditions to ensure that the interface at the base material and cladding material melting, but the matrix material melting too much Resulting in the cladding layer and the matrix element mutual penetration, so that the performance of the cladding layer is affected. Using the simplified formula to calculate the cladding dilution rate, the formula is shown in the formula (1), and the physical quantities in the formula are shown in Figure 2.

$$\eta = \frac{h}{H+h} \times 100\% \tag{1}$$

The dilution rate of the laser cladding layer calculated by formula (1) is shown in Fig. 2. In the process of laser cladding, the energy is dissipated by the three methods of melting the cladding material, the heat transfer of the molten substrate and the substrate. According to the trend of the curve in the figure, it can be seen that the laser power is fixed and the scanning speed is increased, and the dilution rate of the cladding layer shows a decreasing trend. When the scanning speed is constant, the laser power increases and the dilution rate increases.

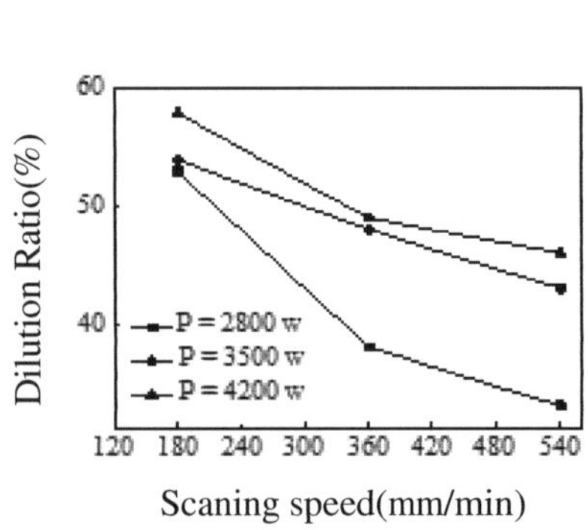

Figure 2. Dilution rate of the cladding layer

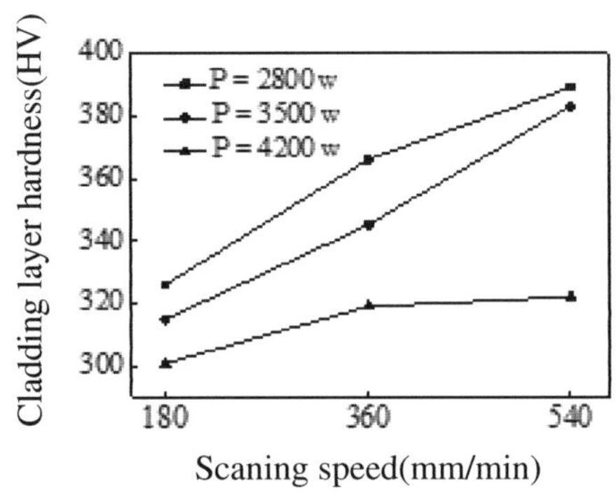

Figure 3. Hardness test of cladding layer

3.3.Microstructure of cladding layer

The surface quality and size of cladding layer prepared by laser cladding can not directly meet the technical requirement of parts. After grinding, it is necessary to pay attention to the upper and middle parts of cladding layer in the study of cladding hardness. This part is the post-processing parts of the surface. The data obtained by averaging the results of the microhardness measurements in the range of 1 mm to 2 mm from the upper surface of the laser clad layer on the single track were used as the hardness test results, hereinafter referred to as the effective hardness of the clad layer. Figure 3 shows the relationship between the microhardness of the cladding layer and the laser power and scanning velocity. When the scanning speed is the same, the effective hardness decreases as the laser power increases. The cladding microstructure under different process parameters is shown in Figure 4 . The grain size increases with the increase of the laser power when the scanning velocity is fixed.

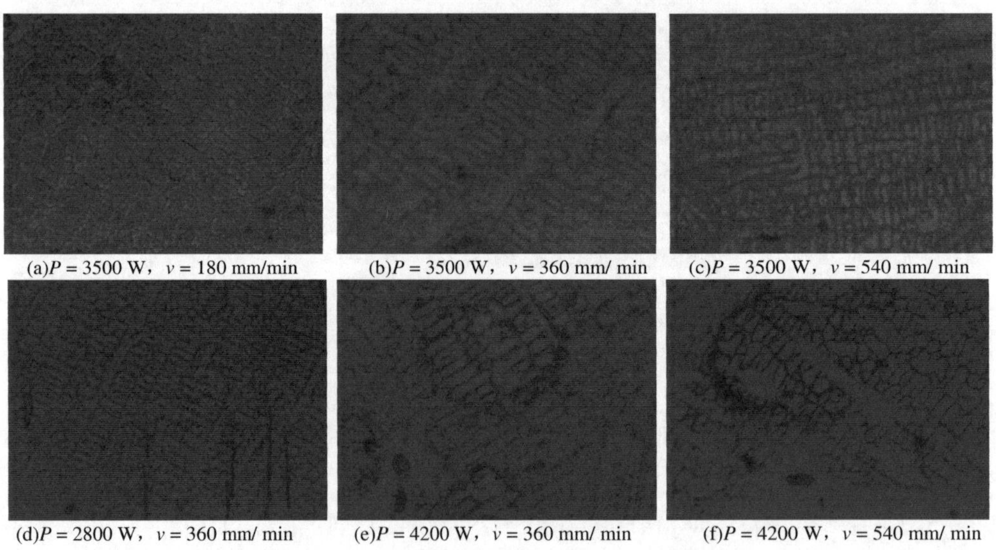

(a)P = 3500 W, v = 180 mm/min (b)P = 3500 W, v = 360 mm/ min (c)P = 3500 W, v = 540 mm/ min

(d)P = 2800 W, v = 360 mm/ min (e)P = 4200 W, v = 360 mm/ min (f)P = 4200 W, v = 540 mm/ min

Figure 4. Cladding layer microstructure

3.4. Friction and Wear Characteristics of Cladding Layer

Wear surface of the workpiece and the surface of the Si3N4 ceramic ball are not smooth, the friction coefficient at the front of the curve in Figure 5 rises rapidly at the beginning of the test. After 400 s, The sample into a more stable wear stage, two pairs of grinding surface tends to run, the friction factor in a certain range of fluctuations. The mean value of the friction coefficient of the Co-based material is 0.77 in the stable wear stage, 0.69 in the 20CrMnTi material, and 20CrMnTi in the abrasion resistance.

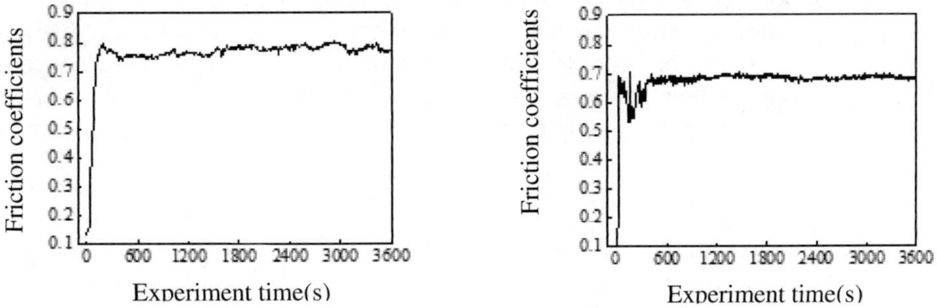

(a) Friction coefficient curve of Co-based clading layer

(b) Friction coefficient curve of 20CrMnTi clading layer

Figure 5. Friction coefficients curve of the clading materials

Co-based material and 20CrMnTi material before and after wear and tear compared to the weight loss shown in Table 1. The wear scar profile of the two materials is similar to that of the two materials, but there is a great difference in the wear depth. The maximum depth of wear trace of the Co-based material is 3.13 μm, and the grinding of 20CrMnTi material The maximum trace depth is 1.45 μm, and the trace of 20CrMnTi material is shallow. According to the studies have shown that by adding Y2O3, CeO2, La2o3, La2O3, La2O3, and La2O3, WC and other strengthening phase can effectively improve the hardness and wear resistance of the alloy cladding layer.

Table 1 Comparison of samples weightlessness before and after wear test

	Co-based cladding	20CrMnTi
Before the test （g）	15.7867	19.4526
After the test （g）	15.7859	19.4523
Sample wear （g）	0.0008	0.0003

3.5.Multilayer laser cladding test

In the cladding process, the laser power to ensure the full melting of the material under the premise of gradual decline, you can avoid the phenomenon of burning collapse. As shown in Figure 6, the laser power was reduced to 800 W when the 11th layer was cladded. The surface of the prepared thin-walled structure was dense and free of hole defects, and the top Smooth without collapse.

Figure 6. Thin-walled structure after multilayer cladding

IOP Conf. Series: Materials Science and Engineering **265** (2017) 012021 doi:10.1088/1757-899X/265/1/012021

4. Conclusion

(1) A large amount of laser cladding experiments were carried out using Co-based alloy powders, and the macroscopic and microscopic characteristics, such as crack sensitivity, dilution rate, microstructure and friction and wear properties, of Co-based cladding were studied. The multi-layer laser cladding experiment of thin-walled structural parts was carried out

(2) High power laser cladding using semiconductor laser can make the thick Co-based cladding material melt and get a flat cladding surface. Co-based cladding has low crack susceptibility, only In the crater occasionally small cracks in the actual processing process should be taken to slow down the power or set the process board to avoid crack defects; Co-based materials need to add hard phase to enhance the wear resistance before they can be applied to the actual remanufacturing repair process in.

(3) The multi-layer cladding strategy with variable power can be used to prepare the non-collapsed thin-walled structural parts. During the cladding process, the laser power can be reduced by ensuring the material is fully melted.

Acknowledgments

This work was supported in part by the people's livelihood Scientific and Technological Programme of QingDao under Grant Nos. 15-9-2-124-nsh.

References

[1] ZHANG J, WU W, and ZHAO L Z 2013 Research progress and development trend of laser cladding *Hot Working Technology* pp 131-134

[2] YANG J X, JIN Y P, and ZHANG N 2016 Present situation and future development of the application of laser cladding technology *MW Metal Forming* pp 13-16

[3] YANG L J 2016 Present situation and future development of the application of laser processing technology *MW Metal Forming* pp 10-12

[4] GU K X, ZHANG H, and WANG H J 2015 Effect of cryogenic treatment on dimensional stability and hardness of 20CrMnTi steel *Heat Treatment of Metal* pp104-107

[5] WANG L, NING Y Q, and QIN L 2015 Development of high power diode laser *Chinese Journal of Lumineescence* pp 1-19

[6] SHEN Y, GYAN Y Q, and BAI S 2011 Development status and prospects of laser cladding technology *Manufacturing Technology & Machine Tool* pp 21-24

[7] VALSECCHI B, PREVITALI B, and VEDANI M 2010 Fiber laser cladding with high content of WC-Co based powder *International Journal of Material Forming* pp 1127-1130

[8] FARNIA A, Malek F, and Ocelík V 2013 Microstructural characterization of Co-based coating deposited by low power pulse laser cladding *Journal of Materials Science,March* pp 2714–2723

IOP Conf. Series: Materials Science and Engineering **265** (2017) 012022 doi:10.1088/1757-899X/265/1/012022

The Normal Condition of Squeeze-film Damping Model for Torsion Micro-Resonators

Pu Li[*] and Zhikang Zhang

School of Mechanical Engineering, Southeast University, Jiangning, Nanjing 211189, People's Republic of China
* seulp@seu.edu.cn

Abstract: The prediction of squeeze-film damping plays a significant role in the design of high Q MEMS devices. This paper presents an analytical solution for the effect of squeeze film damping on a rectangular torsion micro-mirrors whose torsion axis exists in different positions. We defined a variable *l* to represent the distance which torsion axis departs from its central axis. The double sine solution is derived from the linearized Reynolds equation to obtain the pressure distribution under the vibrating plate, and then calculate the analytical expression of the damping constant. Comparing the results with the finite element method(FEM), the accuracy of the model can be verified.

1. Introduction

Squeeze-film damping, the main energy dissipation mechanism of MEMS resonators, plays an important role of device's dynamic response, which have been studied for several decades relating to the torsion micro-mirrors. The common tool named Reynolds equation is used to calculate the squeeze-firm damping [1-2].

With the wide usage of the MEMS device, there is an urgent demand of analytical models describing the dynamic response. Langlois[3] proposed a general equation for the compressible flow gas film, this equation is the traditional Reynolds equation. Bao et al [4] proposed an analytical model for calculating the squeeze film damping of a rectangular torsion mirror at finite tilting angles. Pan et al [5] derived the Fourier series solution and the double sine series solution by solving a linearized Reynolds equation under the assumption of small displacements. However, in most of the previous studies, only the situation that torsion axis existing either in central axis or one side of the microplate have been studied, but lack of the normal situation, in which the torsion axis exists neither the central nor one side of the plate, but in the arbitrary position. In this paper we define *l* as the distance that torsion axis departs from the central axis, and the air pressure under the circular plate is approximated by the double series.

2. Problem formulation

A rigid rectangular plate size $L_x \times L_y$ is shown in Figure 1. *l* is the distance which torsion axis departs

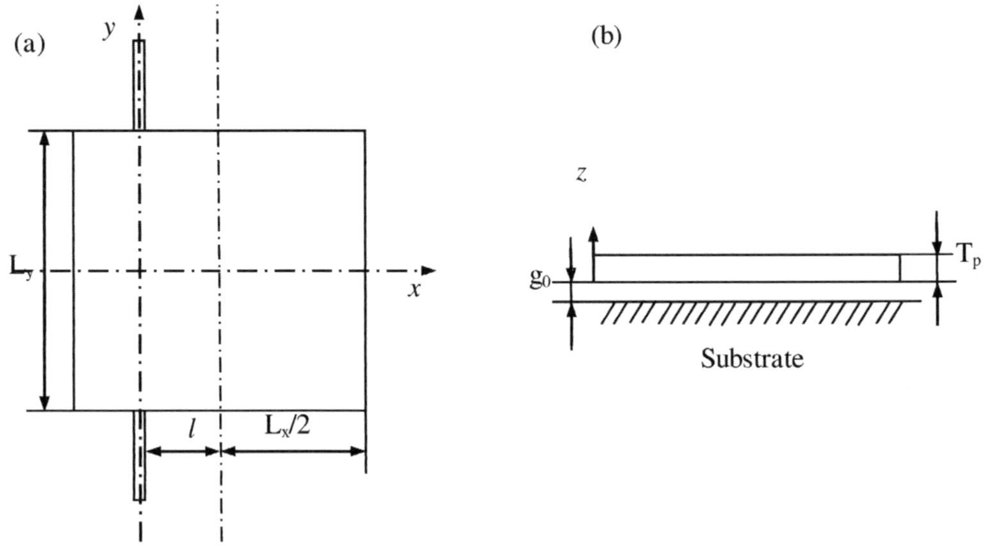

Figure 1. A schematic drawing of a rigid rectangular micro-plate. (a) Top view of perforated microplate. (b) Cross-sectional view

from its central axis, T_p is the thickness of plate, and g_0 the thickness of the gap. In the condition of incompressible gas, the governing equation proposed by Bao et al [1] is

$$\frac{\partial^2 p}{\partial x^2}+\frac{\partial^2 p}{\partial y^2}=\frac{12\mu}{h_0^3}\frac{dh}{dt} \tag{1}$$

Where $p(x,y,t)$ is the air pressure of squeeze-film gap, $p(x,y,t)=p_a+\Delta p(x,y,t)$, p_a is the ambient pressure, Δp the pressure variation, and μ the standard viscosity. $h(t)=g_0+x\theta e^{\omega t}$, in which $\theta e^{\omega t}$ is the angle of the torsion micro-plate.
The following non-dimensional variables are introduced for convenience:

$$P(x,y,t)=\frac{\Delta p(x,y,t)}{p_a},H(t)=h_0 e^{j\omega t},h_0=\frac{\theta_0}{g_0} \tag{2}$$

Substituting equation (2) into equation (1), and linearizing the outcome around p_a and g_0 which leads to

$$\frac{\partial^2 P}{\partial x^2}+\frac{\partial^2 P}{\partial y^2}=\alpha^2\cdot x\frac{dH}{dt} \tag{3}$$

Where $\alpha^2=\frac{12\mu}{p_a g_0^2}$, l is the distance departing from the central axis, and the boundary condition can be changed to

$$P\left(x,-\frac{L_y}{2},t\right)=P\left(x,\frac{L_y}{2},t\right)=P\left(-\frac{L_x}{2}+l,y,t\right)=P\left(\frac{L_x}{2}+l,y,t\right)=0 \tag{4}$$

The expression of $P(x,y,t)$ is able to expressed as a double sine series:

$$P(x,y,t)=\sum_{m=1}^{\infty}\sum_{n=1}^{\infty}a_{mn}\,sin\left[\frac{m\pi}{L_x}\left(x+\frac{L_x}{2}-l\right)\right]sin\left[\frac{n\pi}{L_y}\left(y+\frac{L_y}{2}\right)\right]\cdot e^{j\omega t} \tag{5}$$

Where a_{mn} is the complex amplitude need determination. Obviously, $P(x,y,t)$ satisfies the boundary condition(equation (4)). Substituting equation (5) and $H(t)=h_0e^{j\omega t}$ into equation (3) the following expression can be derived.

$$-\sum_{m=1}^{\infty}\sum_{n=1}^{\infty}\left[\left(\frac{m\pi}{L_x}\right)^2+\left(\frac{n\pi}{L_y}\right)^2\right]a_{mn}sin\left[\frac{m\pi}{L_x}\left(x+\frac{L_x}{2}-l\right)\right]sin\left[\frac{n\pi}{L_y}\left(y+\frac{L_y}{2}\right)\right]=j\omega\alpha^2h_0x \tag{6}$$

According to the orthogonality of trigonometric series, we can lead to

$$\begin{aligned}a_{mn}&=ja_{mn}^I\\&=\frac{8h_0}{mn\pi^2}\left\{L_x-\left[1+(-1)^{m+1}\right]\cdot\left(\frac{L_x}{2}+l\right)\right\}\frac{j\omega\alpha^2}{\left[\left(\frac{m\pi}{L_x}\right)^2+\left(\frac{n\pi}{L_y}\right)^2\right]}\end{aligned} \tag{7}$$

Where a_{mn}^I is the imaginary parts of a_{mn}, and the expression of a_{mn}^I is :

$$a_{mn}^I=\frac{8h_0}{mn\pi^2}\left\{L_x-\left[1+(-1)^{m+1}\right]\cdot\left(\frac{L_x}{2}+l\right)\right\}\frac{\omega\alpha^2}{\left(\frac{m\pi}{L_x}\right)^2+\left(\frac{n\pi}{L_y}\right)^2} \tag{8}$$

$$m=1,2,3...\ ;n=1,3,5...$$

The torque which caused by the pressure of the squeeze film can be derived as

$$T_{squeeze}=\int_{-\frac{L_x}{2}+l}^{\frac{L_x}{2}+l}\int_{-\frac{L_y}{2}}^{\frac{L_y}{2}}\Delta p\cdot xdxdy=jT_{damping}\cdot e^{j\omega t} \tag{9}$$

The total damping torque acting on the microplate are

$$\begin{aligned}T_{damping}=T_{squeeze}^I&=p_a\cdot a_{mn}^I\cdot\int_{-\frac{L_x}{2}+l}^{\frac{L_x}{2}+l}\int_{-\frac{L_y}{2}}^{\frac{L_y}{2}}\sum_{m=1}^{\infty}\sum_{n=1}^{\infty}x\cdot sin\left[\frac{m\pi}{L_x}\left(x+\frac{L_x}{2}-l\right)\right]sin\left[\frac{n\pi}{L_y}\left(y+\frac{L_y}{2}\right)\right]dxdy\\&=-p_ah_0\cdot\sum_{m=1}^{\infty}\sum_{n=1}^{\infty}\frac{16L_xL_y}{m^2n^2\pi^4}\left\{L_x-\left[1+(-1)^{m+1}\right]\cdot\left(\frac{L_x}{2}+l\right)\right\}^2\frac{\omega\alpha^2}{\left(\frac{m\pi}{L_x}\right)^2+\left(\frac{n\pi}{L_y}\right)^2}\end{aligned} \tag{10}$$

Where $T_{squeeze}^I$ is the imaginary part of $T_{squeeze}$.The corresponding damping constant C_θ owing to the pressure of the squeeze film is given by

$$C_\theta=\frac{-T_{damping}}{\theta_0\cdot\omega}=\frac{p_a}{g_0}\sum_{m=1}^{\infty}\sum_{n=1}^{\infty}\frac{16L_xL_y}{m^2n^2\pi^4}\left\{L_x-\left[1+(-1)^{m+1}\right]\cdot\left(\frac{L_x}{2}+l\right)\right\}^2\frac{\alpha^2}{\left(\frac{m\pi}{L_x}\right)^2+\left(\frac{n\pi}{L_y}\right)^2} \tag{11}$$

$$m=1,2,3...\ ;n=1,3,5...$$

3. Validation

In this section, using the present model to calculate the squeeze film damping in a different values of *l*, we will compare the present results with FEM results. In ANSYS, FLUID136 element is used for modelling the squeeze-film. The parameters and dimensions of the rectangular micro-plate are listed in Table 1. Figure 2 shows the comparison of damping constants as a function of distance(*l*) that torsion axis departing from the central axis obtained by the present model (equation (11)).
We varied the value of *l* from 0 to 250 μm, finding that the present results are in good agreement with the FEM results in a wide range of distance. We also find that the damping constant increased as the

increasing of the distance that torsion axis departing from the central axis, and the damping constant also increased as the micro-mirror became more and more large.

Table 1. The parameters and dimensions of the torsion micro-mirror

Parameters	Values	Description
L_x	500 μm	Length of the plate
L_y	500 μm	Width of the plate
T_p	10 μm	Thickness of the plate
g_0	5 μm	Gap spacing
ρ	2330 $\mathrm{Kg/m^3}$	Density
μ_0	1.85×10^{-5} $\mathrm{N\cdot s/m^2}$	Viscosity
f	5000 Hz	Frequency

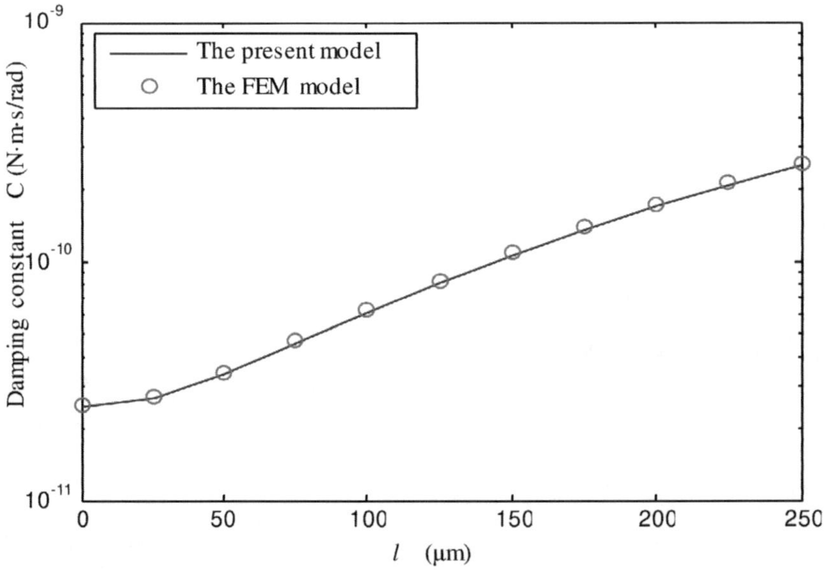

Figure 2. The comparison of damping constants with present model and the FEM model as $L_y = 500\ \mu m$

4. Conclusion

In this paper, an analytical model for squeeze-film damping in a rigid rectangular torsion microplate is presented. The pressure distribution of the gas under the plate calculated by the double sine series, and therefore we get the expressions of damping constant. Comparing them with the FEM results, we find that the analytical model fits the FEM model well, and as the increase of the distance which torsion axis departing from the central axis, the damping constants increased in the same time.

Acknowledgement

This project is supported by National Natural Science Foundation of China (Grant no. 51375091).

Reference

[1] Bao M and Yang H 2007 Squeeze film air damping in MEMS, *Sensors and Actuators A* **136** 3-27

[2] Homentcovschi D and Miles R, 2005 Viscous damping of perforated planar Micromechanical Structures, *Sensors and Actuators A* **119** 544–52

[3] Langlois W E. , 1962 Isothermal squeeze film *Q. Appl. Math*. **20** 131-50
[4] Bao M, Sun Y, J. Zhou and Huang Y, 2006 Squeeze-film air damping of a torsion mirror at a finite tilting angle, *J .Micromech. Microeng*. **16** 2330-35
[5] Pan F Kubby J, Peeters E Tran A and Mukherjee S, 1998 Squeeze film damping effect on the dynamic response of a MEMS torsion mirror, J .Micromech. Microeng. **8** 200-08

Thermoelastic damping in bilayered microbar resonators with circular cross-section

Xiaoyao Liang[1], **Pu Li**[1,*]

[1] School of Mechanical Engineering, Southeast University, Jiangning, Nanjing 211189, People's Republic of China

[*] Corresponding author: seulp@seu.edu.cn

Abstract. It is always a challenge to determine the Thermoelastic damping (TED) in bilayered microbars precisely. In this paper, a model for TED in the bilayered and cantilevered microbar was proposed, in which the total damping was derived by calculating the energy evanished in each layer. The distribution of temperature in the bilayered microbar with a thermodynamically ideal boundary receiving a time-harmonic force is obtained. An infinite summation for the computing of TED in the bilayered slender microbars under axial loading is presented, and the convergence rate of it is discussed. There are little differences between the results computed by our model and that by finite element method (FEM).

1. Introduction

MEMS resonators are used in many fields such as magnetometers, pressure sensor and acoustic sensor. Under most of these circumstances, a high quality factor Q is essential. There exists an important impact of thermoelastic damping (TED) which determines the quality factor Q of resonators on MEMS microstructures. In 1937 and 1938 [1,2], the theory of TED in a vibrating beam was first proposed by Zener, in which a 1-D model was developed which merely considered the heat conduction along the direction of thickness of the microbeam.

Emerging MEMS technologies have validated composite structures which may have significant improvement in performance compared to that of decades ago. Since Zener's model was proposed merely for monolayered beams, precise calculation of TED in composite microstructures is now in great demand for micro-resonators. Bishop and Kinra [3] for the first time reported an overall method to calculate TED in an N-layered structure in 1993. But due to lots of unknown coefficients which depends on the status of microbeams, it is inappropriate to be applied straightly to calculate TED of the microbars and microbeam consisting of composite microstructures.

The bilayered microbars with circular cross-section are as well applied in a lot of MEMS equipment. However, to the best of our knowledge, TED in the oscillation of the slender microbars has not been studied by other researchers. The aim of this paper is to build an analytical model for calculating TED in the bilayered microbars under axial loading. The deduction in this paper is also on the basis of the method proposed by Bishop and Kinra [3].

2. Thermoelastic damping in axially loaded bar-resonators

2.1. The temperature field in a bilayered bar with circular cross-section

Figure 1 shows a bilayered bar with circular cross-section. The bilayered microbar undergoes axial loading. A cylindrical coordinate system (ρ, γ, z) is attached to the left surface of the slender bar. The circular cross-section consists of an inner layer in $0 \le \rho \le r_1$ and an outer layer in $r_1 \le \rho \le r_2$. Apparently, the thicknesses of the inner layer is $R_1 = r_1$ and the thicknesses of the outer layer is $R_2 = (r_2 - r_1)$. d is the length of the bar.

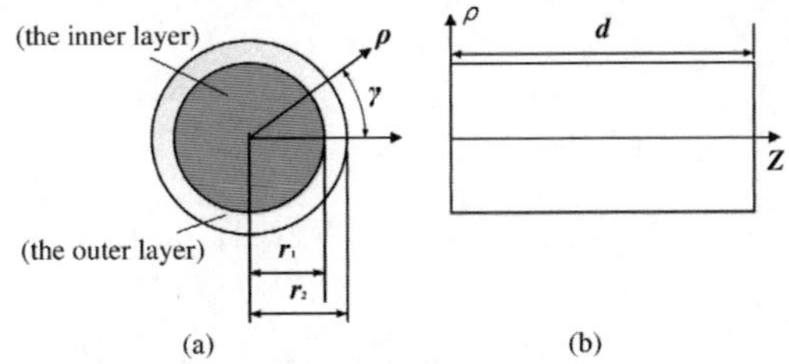

Figure 1. A schematic drawing of a bilayered bar with circular cross-section (a) Cross-section view of the microbar and (b) side view of the microbar

We suppose that the bilayered bar is vulnerable to a time-harmonic uniaxial force with an angular frequency $\hat{\omega}$. The deflection of the bar along the z-direction is

$$w(\rho,\gamma,z,t) = \hat{w}_0(z)\mathrm{e}^{\mathrm{i}\hat{\omega}t} \tag{1}$$

Where $\hat{w}_0(z)$ is the displacement function.

It is assumed that plane section remains plane. Then the strain across the cross-section is uniform whereas stress is piecewise constant. The temperature field in jth layer is

$$\Gamma_j(\rho,\gamma,z,t) = T_j(\rho,\gamma,z,t) - T_0 = \tau_j(\rho,\gamma,z)\mathrm{e}^{\mathrm{i}\hat{\omega}t}, \quad j=1,2 \tag{2}$$

Where j=1,2 denote the inner and the outer layer. $T_j(\rho,\gamma,z,t)$ is the instantaneous temperature and $\Gamma_j(\rho,\gamma,z,t)$ is the variation of temperature about the reference temperature T_0. The constitutive relationships between the strains and stresses is[4]

$$\begin{cases} \varepsilon_{zz,j}(\rho,\gamma,z,t) = \hat{\varepsilon}_{zz,j}(\rho,\gamma,z)\mathrm{e}^{\mathrm{i}\hat{\omega}t} = -\dfrac{\partial \hat{w}_0}{\partial z}\mathrm{e}^{\mathrm{i}\hat{\omega}t} = \left[\dfrac{1}{E_j}\hat{\sigma}_{zz,j}(\rho,\gamma,z)\right]\mathrm{e}^{\mathrm{i}\hat{\omega}t} + \varepsilon_j^{\text{thermal}} \\ \qquad = \left[\dfrac{1}{E_j}\hat{\sigma}_{zz,j}(\rho,\gamma,z) + \hat{\varepsilon}_j^{thermal}\right]\mathrm{e}^{\mathrm{i}\hat{\omega}t} \\ \varepsilon_{\rho\rho,j}(\rho,\gamma,z,t) = \hat{\varepsilon}_{\rho\rho,j}(\rho,\gamma,z)\mathrm{e}^{\mathrm{i}\hat{\omega}t} = \left[-\dfrac{v_j}{E_j}\hat{\sigma}_{zz,j}(\rho,\gamma,z)\right]\mathrm{e}^{\mathrm{i}\hat{\omega}t} + \varepsilon_j^{\text{thermal}} = \left[-v_j\hat{\varepsilon}_{zz,j}(\rho,\gamma,z) + (1+v_j)\hat{\varepsilon}_j^{thermal}\right]\mathrm{e}^{\mathrm{i}\hat{\omega}t} \\ \varepsilon_{\gamma\gamma,j}(\rho,\gamma,z,t) = \hat{\varepsilon}_{\gamma\gamma,j}(\rho,\gamma,z)\mathrm{e}^{\mathrm{i}\hat{\omega}t} = \left[-\dfrac{v_j}{E_j}\hat{\sigma}_{zz,j}(\rho,\gamma,z)\right]\mathrm{e}^{\mathrm{i}\hat{\omega}t} + \varepsilon_j^{\text{thermal}} = \left[-v_m\hat{\varepsilon}_{zz,j}(\rho,\gamma,z) + (1+v_j)\hat{\varepsilon}_j^{thermal}\right]\mathrm{e}^{\mathrm{i}\hat{\omega}t} \end{cases}, \quad j=1,2 \tag{3}$$

Where $\varepsilon_j^{thermal}$ is the thermal strain in the jth layer and the expression of $\varepsilon_j^{thermal}$ is given by

$$\varepsilon_j^{\text{thermal}}(\rho,\gamma,z,t) = \hat{\varepsilon}_j^{thermal}(\rho,\gamma,z)\mathrm{e}^{\mathrm{i}\hat{\omega}t} = \alpha_j\Gamma_j(\rho,\gamma,z,t) = \alpha_j\tau_j(\rho,\gamma,z)\mathrm{e}^{\mathrm{i}\hat{\omega}t} \tag{4}$$

Where α_j is the linear thermal expansion coefficient of the jth layer.

The stress along the z-direction in jth layer can be described as

$$\sigma_{zz,j}(\rho,\gamma,z,t) = \hat{\sigma}_{zz,j}(\rho,\gamma,z)\mathrm{e}^{\mathrm{i}\hat{\omega}t} = -E_j\frac{\partial \hat{w}_0}{\partial z}\mathrm{e}^{\mathrm{i}\hat{\omega}t} \tag{5}$$

The coupled heat conduction equation in the jth layer can be described as

$$\frac{k_j}{C_j}\nabla^2\Gamma_j - \frac{1}{C_j}\frac{E_j\alpha_j T_0}{1-2v_j}\frac{\partial \varepsilon_j}{\partial t} = \frac{\partial \Gamma_j}{\partial t} \qquad j=1,2 \tag{6}$$

Where k_j is the thermal conductivity in the jth layer, C_j is the specific heat per unit volume in the jth layer, ε_j is the cubical dilation in the jth layer. The expression of ε_j is given by

$$\varepsilon_j = \varepsilon_{zz,j}(\rho,\gamma,z,t) + \varepsilon_{\rho\rho,j}(\rho,\gamma,z,t) + \varepsilon_{\gamma\gamma,j}(\rho,\gamma,z,t) = \left[(1-2v_j)\hat{\varepsilon}_{zz,j}(\rho,\gamma,z) + 2(1+v_j)\alpha_j\tau_j\right]\mathrm{e}^{\mathrm{i}\hat{\omega}t} \tag{7}$$

The temperature field in the bilayered bar under axial loading is axisymmetric and in the long slender bar, the temperature gradients along the ρ-direction are much greater than those along the z-and

γ-direction, thus $\nabla^2\Gamma_j = \frac{1}{\rho}\frac{\partial}{\partial\rho}(\rho\frac{\partial\Gamma_j}{\partial\rho}) + \frac{1}{\rho^2}\frac{\partial^2\Gamma_j}{\partial\gamma^2} + \frac{\partial^2\Gamma_j}{\partial z^2} \approx \frac{1}{\rho}\frac{\partial}{\partial\rho}(\rho\frac{\partial\Gamma_j}{\partial\rho})$. Substituting equation (7) into equation (6), and using $2\hat{\Delta}_{Zener,j}\frac{1+v_j}{1-2v_j} << 1$,we obtain

$$\frac{k_j}{C_j}\left(\frac{\partial^2\Gamma_j}{\partial\rho^2} + \frac{\partial\Gamma_j}{\rho\partial\rho}\right) + \frac{\hat{\Delta}_{zener,j}}{\alpha_j} i\hat{\omega}\frac{\partial\hat{w}_0}{\partial z}e^{i\hat{\omega}t} = \frac{\partial\Gamma_j}{\partial t} \tag{8}$$

Where $\hat{\Delta}_{Zener,j} = \frac{E_j\alpha_j^2 T_0}{C_j}$ is the Zener's modulus in the *j*th layer. Table 1 lists the values of Zener's moduli for common materials at T_0=300K.

Table 1. Mechanical and thermal properties of the materials at T_0=300K.

Material	E(GPa)	P_0(Kg m^{-3})	k(W m^{-1}K^1)	α(K^{-1})	C(J m^{-3} K^{-1})	υ	Δ_{Zener}
Si	160	2300	150	2.6 x 10^{-6}	1.6 x 10^6	0.22	2 x 10^{-4}
SiO_2	70	2200	1.2	0.5 x 10^{-6}	1.5 x 10^6	0.18	3.5 x 10^{-6}

For the bilayered bar, the temperature at $\rho=0$ is finite, the inner layer and the outer layer at $\rho=r_1$ are in perfect thermal contact, and the surface of the outer layer at $\rho=r_2$ is adiabatic. Thus the boundary conditions are as follows.

$$\begin{cases} \Gamma_1(\rho,t) = finite & \rho=0\ ,\ \Gamma_1(\rho,t)=\Gamma_2(\rho,t) & \rho=r_1 \\ k_1\frac{\partial\Gamma_1(\rho,t)}{\partial\rho} = k_2\frac{\partial\Gamma_2(\rho,t)}{\partial\rho} & \rho=r_1\ ,\ \frac{\partial\Gamma_2(\rho,t)}{\partial\rho}=0 & \rho=r_2 \end{cases} \tag{9}$$

To obtain the temperature field in composite structures of microbars, a new method which was based on the trigonometric series was proposed by Ozisik [5] in 1980. Based on the Ozisik's theory, the distribution of temperature $\Gamma_j(\rho,t)$ in this paper could be described as

$$\Gamma_1(\rho,t) = \tau_1(\rho)e^{i\hat{\omega}t} = \sum_{n=1}^{\infty} g_n\phi_{1,n}(\rho)e^{i\hat{\omega}t},\quad 0\le\rho\le r_1 \tag{10}$$

$$\Gamma_2(\rho,t) = \tau_2(\rho)e^{i\hat{\omega}t} = \sum_{n=1}^{\infty} g_n\phi_{2,n}(\rho)e^{i\hat{\omega}t},\quad r_1\le\rho\le r_2 \tag{11}$$

Where using the trigonometric series, $\phi_{j,n}(\rho)$ in the *j*th layer can be described as

$$\phi_{j,n}(\rho) = A_{j,n}J_0\left(\frac{\Omega_n}{\sqrt{\zeta_j}}\rho\right) + B_{j,n}Y_0\left(\frac{\Omega_n}{\sqrt{\zeta_j}}\rho\right) \tag{12}$$

Where $\Omega_n, A_{j,n}, B_{j,n}$ *and* g_n are the undetermined real coefficients, Ω_n is the *n*th eigenvalue, and $\zeta_j = \frac{k_j}{C_j}$. $J_i\left(\frac{\Omega_n}{\sqrt{\zeta_j}}\rho\right) and Y_i\left(\frac{\Omega_n}{\sqrt{\zeta_j}}\rho\right)$ are the *i*th order Bessel function in *j*th layer. Obviously , the boundary condition at ρ =0 requires $B_{1,n}$=0, therefore $\phi_{1,n}$ and $\phi_{2,n}$ are reduced to [4]

$$\phi_{1,n}(\rho) = J_0\left(\frac{\Omega_n}{\sqrt{\zeta_1}}\rho\right) \qquad 0\le\rho<r_1 \tag{13}$$

$$\phi_{2,n}(\rho) = A_{2,n}J_0\left(\frac{\Omega_n}{\sqrt{\zeta_2}}\rho\right) + B_{2,n}Y_0\left(\frac{\Omega_n}{\sqrt{\zeta_2}}\rho\right) \qquad r_1\le\rho<r_2 \tag{14}$$

2.1.1. Determination of $\Omega_n, A_{2,n}$ *and* $B_{2,n}$. Substituting equations (10) and (11) into (9), these can be described in the matrix form as

$$\begin{bmatrix} J_0(\psi) & -J_0\left(\frac{r_1}{r_2}\eta\right) & -Y_0\left(\frac{r_1}{r_2}\eta\right) \\ KJ_1(\psi) & -J_1\left(\frac{r_1}{r_2}\eta\right) & -Y_1\left(\frac{r_1}{r_2}\eta\right) \\ 0 & J_1(\eta) & Y_1(\eta) \end{bmatrix} \begin{bmatrix} 1 \\ A_{2,n} \\ B_{2,n} \end{bmatrix} = \begin{bmatrix} 0 \\ 0 \\ 0 \end{bmatrix} \quad (15)$$

Where $\psi = \frac{r_1 \Omega_n}{\sqrt{\zeta_1}}$ $\eta = \frac{r_2 \Omega_n}{\sqrt{\zeta_2}}$ $K = \frac{k_1}{k_2}\sqrt{\frac{\zeta_2}{\zeta_1}}$

The solution of equation (15) is

$$\begin{cases} A_{2,n} = \left[J_0(\psi)Y_1\left(\frac{r_1}{r_2}\eta\right) - KJ_1(\psi)Y_0\left(\frac{r_1}{r_2}\eta\right)\right] \times \left[J_0\left(\frac{r_1}{r_2}\eta\right)Y_1\left(\frac{r_1}{r_2}\eta\right) - J_1\left(\frac{r_1}{r_2}\eta\right)Y_0\left(\frac{r_1}{r_2}\eta\right)\right]^{-1} \\ B_{2,n} = \left[KJ_1(\psi)J_0\left(\frac{r_1}{r_2}\eta\right) - J_0(\psi)J_1\left(\frac{r_1}{r_2}\eta\right)\right] \times \left[J_0\left(\frac{r_1}{r_2}\eta\right)Y_1\left(\frac{r_1}{r_2}\eta\right) - J_1\left(\frac{r_1}{r_2}\eta\right)Y_0\left(\frac{r_1}{r_2}\eta\right)\right]^{-1} \end{cases} \quad (16)$$

Obviously $A_{2,n}$ and $B_{2,n}$ are nonzero and real constants. This situation leads to the following equation for the solution of Ω_n

$$\begin{vmatrix} J_0(\psi) & -J_0\left(\frac{r_1}{r_2}\eta\right) & -Y_0\left(\frac{r_1}{r_2}\eta\right) \\ KJ_1(\psi) & -J_1\left(\frac{r_1}{r_2}\eta\right) & -Y_1\left(\frac{r_1}{r_2}\eta\right) \\ 0 & J_1(\eta) & Y_1(\eta) \end{vmatrix} = 0 \quad (17)$$

The eigenvalues $\Omega_1 < \Omega_2 < \cdots < \Omega_n < \cdots$ are deduced from equation (17).

2.1.2. Determination of g_n. Substituting equations (10) and (13) into equation (8). Multiplying both sides of obtaining equation by $\rho\phi_{1,n}(\rho)$, and integrating the results with respect to ρ from 0 to r_1, lead to

$$g_n H_{1,n} + J_{I,n}\frac{E_1\alpha_1 T_0}{k_1} i\hat{\omega}\frac{\partial \hat{w}_0}{\partial z} = g_n \frac{J_{III,n}}{\zeta_1} i\hat{\omega} \quad (18)$$

Substituting equations (11) and (14) into equation (8). Multiplying both sides of obtaining equation by $\rho\phi_{2,n}(\rho)$, and integrating the results with respect to ρ from r_1 to r_2, lead to

$$g_n H_{2,n} + J_{II,n}\frac{E_2\alpha_2 T_0}{k_2} i\hat{\omega}\frac{\partial \hat{w}_0}{\partial z} = g_n \frac{J_{IV,n}}{\zeta_2} i\hat{\omega} \quad (19)$$

where $H_{1,n} = \int_0^{r_1} \rho\phi_{1,n}(\rho)\left(\frac{\partial^2 \phi_{1,n}}{\partial \rho^2} + \frac{\partial \phi_{1,n}}{\rho \partial \rho}\right) d\rho = -\left(\frac{\Omega_n}{\sqrt{\zeta_1}}\right)^2 J_{III,n}$, $H_{2,n} = \int_{r_1}^{r_2} \rho\phi_{2,n}(\rho)\left(\frac{\partial^2 \phi_{2,n}}{\partial \rho^2} + \frac{\partial \phi_{2,n}}{\rho \partial \rho}\right) d\rho = -\left(\frac{\Omega_n}{\sqrt{\zeta_2}}\right)^2 J_{IV,n}$.

Using equations (18) and (19), we obtain

$$g_n = \frac{\partial \hat{w}_0}{\partial z} T_0 \frac{i\hat{\omega}}{i\hat{\omega} + \Omega_n^2} \frac{E_1\alpha_1 J_{I,n} + E_2\alpha_2 J_{II,n}}{C_1 J_{III,n} + C_2 J_{IV,n}} \quad (20)$$

Where

$$J_{I,n} = \int_0^{r_1} \rho' \phi_{1,n}(\rho') d\rho' = \frac{r_1\sqrt{\zeta_1}}{\Omega_n} J_1\left(\frac{\Omega_n}{\sqrt{\zeta_1}} r_1\right)$$

$$J_{II,n} = \int_{r_1}^{r_2} \rho' \phi_{2,n}(\rho') d\rho' = \frac{\sqrt{\zeta_2}}{\Omega_n}\left[-r_1 A_{2,n} J_1\left(\frac{\Omega_n}{\sqrt{\zeta_2}} r_1\right) - r_1 B_{2,n} Y_1\left(\frac{\Omega_n}{\sqrt{\zeta_2}} r_1\right) + r_2 A_{2,n} J_1\left(\frac{\Omega_n}{\sqrt{\zeta_2}} r_2\right) + r_2 B_{2,n} Y_1\left(\frac{\Omega_n}{\sqrt{\zeta_2}} r_2\right)\right]$$

$$J_{III,n}=\int_0^{r_1}\rho'\phi_{1,n}{}^2(\rho')d\rho'=\frac{1}{2}r_1^2[J_0{}^2(\frac{\Omega_n}{\sqrt{\zeta_1}}r_1)+J_1{}^2(\frac{\Omega_n}{\sqrt{\zeta_1}}r_1)]$$

$$J_{IV,n}=\int_{r_1}^{r_2}\rho'\phi_{2,n}{}^2(\rho')d\rho'=\frac{1}{2}r_2^2\{A_{2,n}{}^2J_0{}^2(\frac{\Omega_n}{\sqrt{\zeta_2}}r_2)+2A_{2,n}B_{2,n}J_0(\frac{\Omega_n}{\sqrt{\zeta_2}}r_2)Y_0(\frac{\Omega_n}{\sqrt{\zeta_2}}r_2)+[A_{2,n}J_1(\frac{\Omega_n}{\sqrt{\zeta_2}}r_2)+B_{2,n}Y_1(\frac{\Omega_n}{\sqrt{\zeta_2}}r_2)]^2+B_{2,n}{}^2Y_0{}^2(\frac{\Omega_n}{\sqrt{\zeta_2}}r_2)\}$$
$$-\frac{1}{2}r_1^2\{A_{2,n}{}^2J_0{}^2(\frac{\Omega_n}{\sqrt{\zeta_2}}r_1)+2A_{2,n}B_{2,n}J_0(\frac{\Omega_n}{\sqrt{\zeta_2}}r_1)Y_0(\frac{\Omega_n}{\sqrt{\zeta_2}}r_1)+[A_{2,n}J_1(\frac{\Omega_n}{\sqrt{\zeta_2}}r_1)+B_{2,n}Y_1(\frac{\Omega_n}{\sqrt{\zeta_2}}r_1)]^2+B_{2,n}{}^2Y_0{}^2(\frac{\Omega_n}{\sqrt{\zeta_2}}r_1)\}$$

Substituting equations (13), (14) and (20) into equations (10) and (11), the temperature fields in the inner layer and outer layer are given by

$$\Gamma_1(\rho,t)=\tau_1(\rho)\mathrm{e}^{i\hat{\omega}t}=T_0\frac{\partial\hat{w}_0}{\partial z}\sum_{n=1}^{\infty}\frac{E_1\alpha_1J_{I,n}+E_2\alpha_2J_{II,n}}{C_1J_{III,n}+C_2J_{IV,n}}\frac{i\hat{\omega}}{\Omega_n^2+i\hat{\omega}}\phi_{1,n}(\rho)e^{i\hat{\omega}t} \tag{21}$$

$$\Gamma_2(\rho,t)=\tau_2(\rho)\mathrm{e}^{i\hat{\omega}t}=T_0\frac{\partial\hat{w}_0}{\partial z}\sum_{n=1}^{\infty}\frac{E_1\alpha_1J_{I,n}+E_2\alpha_2J_{II,n}}{C_1J_{III,n}+C_2J_{IV,n}}\frac{i\hat{\omega}}{\Omega_n^2+i\hat{\omega}}\phi_{2,n}(\rho)e^{i\hat{\omega}t} \tag{22}$$

2.2 Thermoelastic damping in a bilayered bar with circular cross-section

On the basis of the method derived by Bishop and Kinra [3], TED in the slender and bilayered bar is described as

$$Q^{-1}=\frac{1}{2\pi}\frac{\Delta\hat{W}}{\hat{W}}=\frac{1}{2\pi}\frac{\Delta\hat{W}_1+\Delta\hat{W}_2}{\hat{W}_1+\hat{W}_2} \tag{23}$$

Where $\hat{W}_1$ and $\hat{W}_2$ are the maximum reserved energies in the inner and outer layers, $\Delta\hat{W}_1$ and $\Delta\hat{W}_2$ are the energy evanished per period of vibration in the two layers.

The energy evanished can be described as

$$\Delta\hat{W}=\Delta\hat{W}_1+\Delta\hat{W}_2=\iiint_{V_1}\pi E_1\frac{\partial\hat{w}_0}{\partial z}\mathrm{Im}(\alpha_1\tau_1)dV_1+\iiint_{V_2}\pi E_2\frac{\partial\hat{w}_0}{\partial z}\mathrm{Im}(\alpha_2\tau_2)dV_2$$
$$=2\pi^2T_0\int_{z=0}^{d}\left(\frac{\partial\hat{w}_0}{\partial z}\right)^2\sum_{n=1}^{\infty}\frac{\Omega_n^2\hat{\omega}}{\Omega_n^4+\hat{\omega}^2}\frac{(E_1\alpha_1J_{I,n}+E_2\alpha_2J_{II,n})^2}{C_1J_{III,n}+C_2J_{IV,n}} \tag{24}$$

The maximum strain energy reserved $\hat{w}$ can be described as [3]

$$\hat{W}=\hat{W}_1+\hat{W}_2=\frac{1}{2}\iiint_{V_1}\sigma_{zz}\varepsilon_{zz}dV_1+\frac{1}{2}\iiint_{V_2}\sigma_{zz}\varepsilon_{zz}dV_2=\frac{1}{2}\pi[E_1r_1^2+E_2(r_2^2-r_1^2)]\int_{z=0}^{d}\left(\frac{\partial\hat{w}_0}{\partial z}\right)^2dz \tag{25}$$

Substituting equations (24) and (25) into equation (23), we obtain

$$Q^{-1}=\frac{1}{2\pi}\frac{\Delta\hat{W}}{\hat{W}}=\frac{1}{2\pi}\frac{2\pi^2\int_{z=0}^{d}\left(\frac{\partial\hat{w}_0}{\partial z}\right)^2dz\sum_{n=1}^{\infty}\frac{\Omega_n^2\hat{\omega}}{\Omega_n^4+\hat{\omega}^2}p_n}{\frac{1}{2}\pi\int_{z=0}^{d}\left(\frac{\partial\hat{w}_0}{\partial z}\right)^2dz[E_1r_1^2+E_2(r_2^2-r_1^2)]}=2\frac{\sum_{n=1}^{\infty}\frac{\Omega_n^2\hat{\omega}}{\Omega_n^4+\hat{\omega}^2}p_n}{E_1r_1^2+E_2(r_2^2-r_1^2)} \tag{26}$$

Where $p_n=T_0\frac{(E_1\alpha_1J_{I,n}+E_2\alpha_2J_{II,n})^2}{C_1J_{III,n}+C_2J_{IV,n}}$

3. validation and discussions

Here the present model is simplified to fit a monolayered microbar to verify the validity of itself. A detailed deduction is as follow that demonstrates our model (equation (26)) could reduce to the monolayered model under the circumstance of $E_1=E_2=E$, $C_1=C_2=C$, $\alpha_1=\alpha_2=\alpha$, $k_1=k_2=k$ and $v_1=v_2=v$. Careful observation of equation (15), (16) and (26), we obtain

$$Q^{-1}=2T_0\frac{\sum\frac{\Omega_n^2\hat{\omega}}{\Omega_n^4+\hat{\omega}^2}}{Er_2^2}\frac{E^2\alpha^2J_1^2(\frac{\Omega_n}{\sqrt{\zeta}}r_2)}{C(J_{III,n}+J_{IV,n})}=0 \tag{27}$$

In this case, thermoelastic damping will be identically zero. i.e., provided that two kinds of materials in one structure have the same temperature change at given $\hat{w}_0(z)$, then there would be no differences in temperature, no interchange of heat between the two layers, therefore no damping [3].

3.1. Comparison with the finite element method (FEM)

The infinite summations in our model (equation (26)) for TED are calculated on MATLAB. The value of p_n determines the convergence rate of equation (26). In this subsection, we truncate the series (equation (26)) to the finite terms. The convergence rate for a slender and bilayered bar is shown in Figure 2. A total of ten terms were enough to accomplish convergence for the reason that the outcome with n=10 are identical to that with n=50.

Figure 3 shows the values of p_n/S as a function of n. Obviously, the sum of the top ten items of p_n/S is 0.974 that is much larger than $p_{11}/S, p_{12}/S, p_{13}/S...$. So, a total of ten terms were sufficient to achieve convergence. Where S is the infinite summation of p_n.

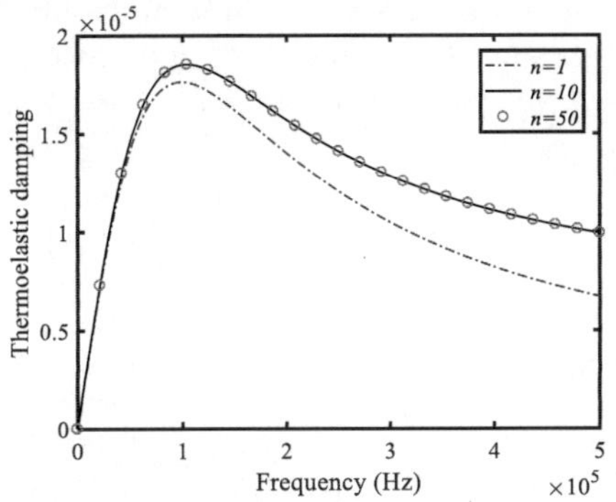

Figure 2. Convergence analysis of the present Model in Si/SiO$_2$ with r_1=8 μm, r_2=10 μm .

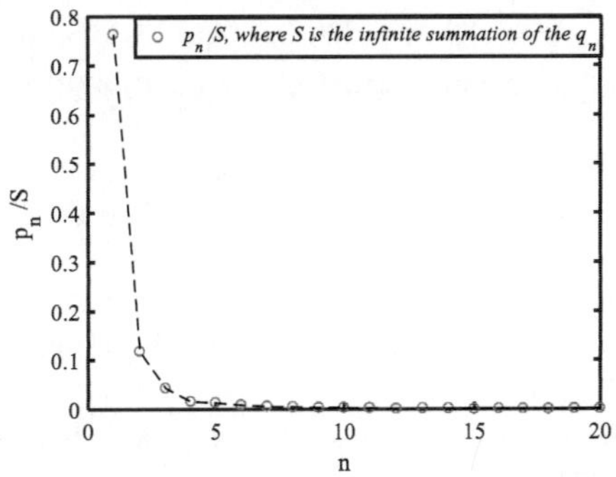

Figure 3. The value of p_n/S as a function of n in the Si/SiO$_2$ with r_1=8 μm, r_2=10 μm.

Now, we use present model (equation (26)) to compare these predictions with the FEM results (ANSYS). In ANSYS, the microbars are excited by an axial force $F_0 Sin(\hat{\omega} t)$, where F_0=0.1 Mpa. In this case, a structure-thermal harmonic analysis was applied to calculate TED in the slender and bilayered microbar, which was meshed with solid 227 elements. The lengths of the bilayered microbar are fixed at d=300μm. The thickness of the inner layer (Si) and the outer layer (SiO$_2$) of the bilayered microbar are R_1=8 μm and R_2=2 μm, respectively. Figure 4 shows the TED of the microbar with r_1=8μm and r_2=10μm computed by the FEM method and present model. Apparently, the outcome calculated by our model also match very well with that by FEM in these situations.

3.2. The Optimization of Si bars

Figure 5 shows the present model results for bilayered microbars with $d=300\mu m$ and $r_2=10\mu m$ at r_1/r_2=0.8, 0.85, 0.9 and 0.95, respectively. The structure of bilayered microbars is Si/SiO$_2$. Under these circumstances, favorable convergence was accomplished by extracting no more than 10 terms. Apparently, in the case of r_1/r_2=0.95, the bilayered microbar has the smallest value of peak because bilayered microbars are more close to the monolayered microbars (Si).

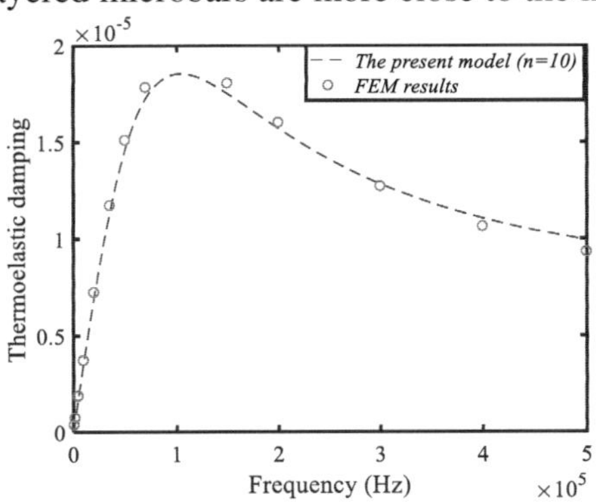

Figure 4. Comparison of TED obtained by the present model and the ANSYS model.

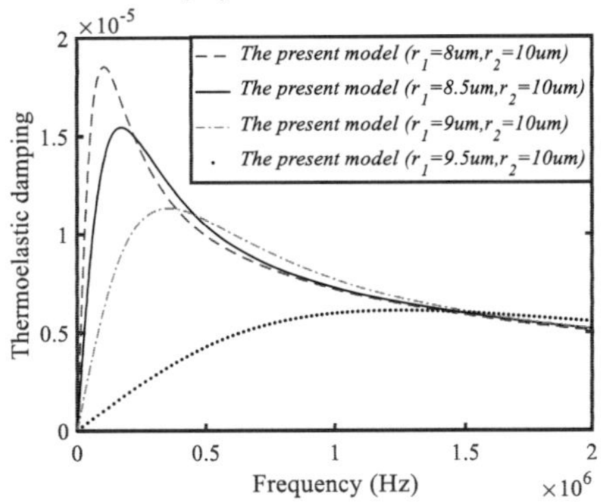

Figure 5. Comparison of TED obtained by the ANSYS model four cantilever bars. The four bars have different thickness of the outer layer.

4. Conclusions

It is always a challenge to determine the TED in bilayered microbars precisely. This paper proposed a precise model for calculating TED in the slender and bilayered and cantilevered microbar, in which the total damping was derived by calculating the energy evanished in each layer.

TED in the bilayered microbars can be predicted by equation (26). The present model under the circumstance of n=10 does succeed to provide a precise forecast for TED in bilayered microbars. The results computed by our present model matches very well with that by FEM. Our model shows that the peak damping in the case of r_1/r_2 =0.95 exhibit lower value than those of other bars.

This model is useful merely for the slender bar with circular cross-section. In this situation, the thermal gradients along the bar radius orientation are far greater than that along the z- and y-orientations. The temperature gradients along the z- and y- directions could be neglected.

Acknowledgment

This project is supported by National Natural Science Foundation of China (Grant no. 51375091).

References

[1] Zener C 1937 Internal friction in solids I. Theory of internal friction in reeds *Phys. Rev.* **52** 230-5
[2] Zener C 1938 Internal friction in solids II. General theory of thermoelastic internal friction *Phys. Rev.* **53** 90-9
[3] Bishop JE and Kinra VK 1993 Thermoelastic damping of a laminated beam in flexure and extension *J. Reinf. Plast. Compos.* **12** 210-26
[4] Parkus H 1976 *Thermoelasticity* (New York: Springer-Verlag)
[5] Ozisik MN 1980 *Heat Conduction* (New York: John Wiley & Sons)

IOP Conf. Series: Materials Science and Engineering **265** (2017) 012024 doi:10.1088/1757-899X/265/1/012024

Research on Creep Relaxation Non-uniformity and Effect on Performance of Combined Rotor

LIU Qingya[1,2], **He Jingfei** [1,2], **Zhao Lijia**[1,2]

[1] College of Mechanical and Electrical Engineering, Central South University, Changsha, Yue Lu Street, China

[2] Key Laboratory of High Performance and Complex Manufacturing, Central South University, Changsha, Yue Lu Street, China

Abstract: The combined rotor of gas turbine is connected by a certain number of rod bolts. It works in the high temperature environment for a long time, and the rod bolts will creep and relax. Under the influence of elastic interaction, the loss of pretightening force of rod bolts at different positions is non-uniform, which will cause the connection of the combined rotor to be out of tune. In this paper, the creep relaxation non-uniformity model for a class F heavy duty gas turbine is established. On the basis of this, the performance degradation and structural strength change of combined rotor resulting from creep relaxation non-uniformity of rod bolts are studied. The results show that the ratio of preload mistuning increases with time and then converges, and there is a threshold inflection point in about seven thousand hours.

1. Introduction

The combined rotor of gas turbine is divided into three parts: compressor, connecting section and turbine. The compressor is composed of 17 stage discs, and is compressed by 12 long rod bolts. The turbine section is composed of 5 stage discs, and is clamped by 12 short rod bolts, as shown in Figure 1. Combined rotor of gas turbine serve in high temperature, high pressure and high speed conditions, due to the impact of micro crack [1] [2] [3] , rod unwinding relaxation [4], creep relaxation of rod [5], slip of discs [6] and other factors, which will cause performance degradation and structural strength reduction of combined rotor .Under high temperature conditions, the creep of rod bolts will cause the reduction of preload, resulting in lower connection stiffness and severe vibration of rotor system. Therefore, it is important to study the creep relaxation of rod bolts.

Some scholars have studied the creep relaxation of rotor system at high temperature. Y Sun [7] studied the creep characteristics of the rotor under bending residual stress. It was found that the creep relaxation alone was not enough to lead to the bending of the rotor during the start-up stage. Y Su [8] carried out the stiffness equivalent treatment on the contact interface of combined rotor. The results show that the creep relaxation of the rod bolts will cause the natural frequency drift of the combined rotor.

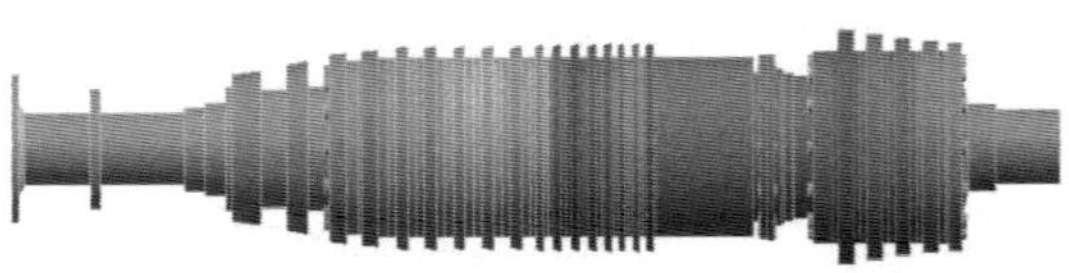

Figure 1. structure diagram of combined rotor

Figure 2. Sketch of rod bolt set

H Zhang [9] analyzed the creep of the rotor disk. It was found that the creep rate of the 1 stage disk is the fastest, and the more the number of rod bolts, the greater the frequency drift. H Liu [10] studied the effect of creep on the fatigue of turbine rotors. It was found that the creep changed the steady strain of the rotor.

To sum up, scholars have studied the creep relaxation related mechanism of the rotor and single rod bolt, but there is little research on the creep relaxation mechanism of rod bolts group. Therefore, in this paper, rod bolts group of a gas turbine is taken as the object of study, and the influence factors of creep relaxation non-uniformity of the bolts are analyzed.

2. Mechanism of Creep Relaxation Non-uniformity of Rod Bolts

To ensure the relative uniformity of the force, the two cycle pre tightening method is adopted in the actual loading stage, and two rod bolts in the diagonal position stretch simultaneously. In order to describe the influence of the interaction between the rod bolts in different positions of the combined rotor, the elastic interaction stiffness Km,n is introduced .The interaction stiffness Km,n indicates that when the rod bolt n is stretched, the disc at the rod bolt m has the capability of resisting the elastic deformation, as shown in Figure 2.

After completion of loading, the preload variation can be expressed as:

$$\begin{cases} \Delta F_{1,3} + \Delta F_{1,9} = -F(K_{1,3}^{-1} + K_{1,9}^{-1})(K_{b1}^{-1} + K_{1,1}^{-1})^{-1} \\ \Delta F_{4,3} + \Delta F_{4,9} = -F(K_{4,3}^{-1} + K_{4,9}^{-1})(K_{b4}^{-1} + K_{4,4}^{-1})^{-1} \\ \Delta F_{2,3} + \Delta F_{2,9} = -F(K_{2,3}^{-1} + K_{2,9}^{-1})(K_{b2}^{-1} + K_{2,2}^{-1})^{-1} \\ \Delta F_{5,3} + \Delta F_{5,9} = -F(K_{5,3}^{-1} + K_{5,9}^{-1})(K_{b5}^{-1} + K_{5,5}^{-1})^{-1} \\ \Delta F_{6,3} + \Delta F_{6,9} = -F(K_{6,3}^{-1} + K_{6,9}^{-1})(K_{b6}^{-1} + K_{6,6}^{-1})^{-1} \end{cases} \tag{1}$$

The general expression of the high temperature creep model is:

$$\delta_{bcr} = L_b C_1 \sigma^{c_2} t^{1+c_3} e^{-\frac{c_4}{T}} / 1 + C_3 \tag{2}$$

According to the deformation coordination of the contact surface, the change of the contact position is consistent with the creep extension of the rod bolts. The general model of creep relaxation Non-uniformity is as follows: $\Delta F^c = -K^{-1}\delta_{bcr}$ (3)

Where

$$K = \begin{bmatrix} K_{1,1}^{-1} & K_{1,2}^{-1} & K_{1,3}^{-1} & \cdot\cdot\cdot & K_{1,12}^{-1} \\ K_{2,1}^{-1} & K_{2,2}^{-1} & K_{2,3}^{-1} & \cdot\cdot\cdot & K_{2,12}^{-1} \\ K_{3,1}^{-1} & K_{3,2}^{-1} & K_{3,3}^{-1} & \cdot\cdot\cdot & K_{3,12}^{-1} \\ \cdot & \cdot & \cdot & & \cdot \\ \cdot & \cdot & \cdot & \cdot\cdot\cdot & \cdot \\ \cdot & \cdot & \cdot & & \cdot \\ K_{12,1}^{-1} & K_{12,2}^{-} & K_{12,3}^{-1} & \cdot\cdot\cdot & K_{12,12}^{-1} \end{bmatrix}$$

3. Creep Relaxation Non-uniformity Analysis of Rod Bolts

3.1 The Mistuning Ratio Under Different Initial Preload Conditions

When the initial pretension is exactly the same, because of the elastic interaction between the bolts at different positions, the variation of preload will be uneven, which will cause the rotor to be out of tune. The ratio of preload mistuning of rod bolt set is changed [11], as shown in Figure 3. The ratio of preload mistuning of rod bolt set is 1.19% at the end of the simulation.

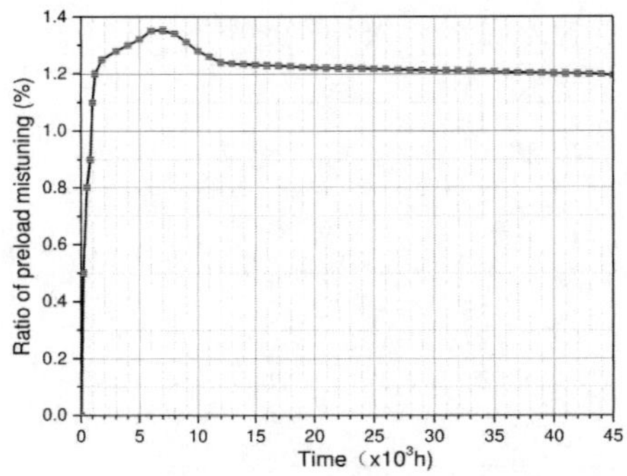

Figure 3. When the initial preload is tuned, mistuning ratio changes with time

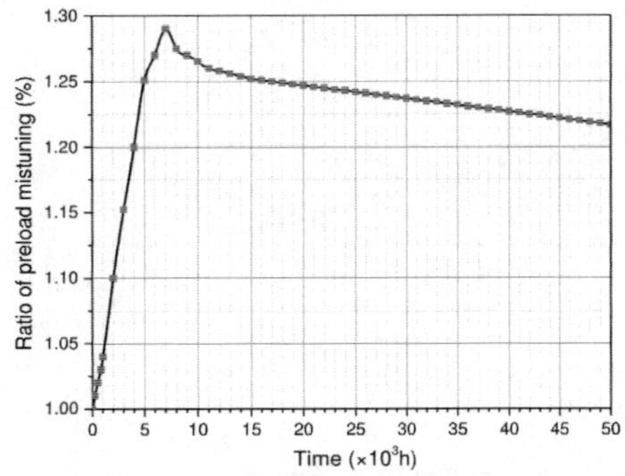

Figure 4. Mistuning ratio versus time under the condition of initial mistuning ratio of 1%

According to the formula of elastic interaction at the pre tightening stage, the ratio of initial pre load is less than 1%. Therefore, the simulation of creep relaxation is carried out with the initial mistuning ratio of 1%, and the simulation time is 50,000 hours, as shown in the Figure 4. The mistuning ratio first diverges and then converges, and there is a threshold for the mistuning curve of rod bolts. When the combined rotor works for 7000 hours, the mistuning threshold is 1.29%.

During the assembly process, the initial mistuning ratio is limited to less than 5%. Under the condition of initial mistuning ratio of 5%, mistuning ratio changes with time, as shown in the Figure 5. At the end of the 50000th hour, the mistuning ratio of the rod bolts is reduced to 1.18%.

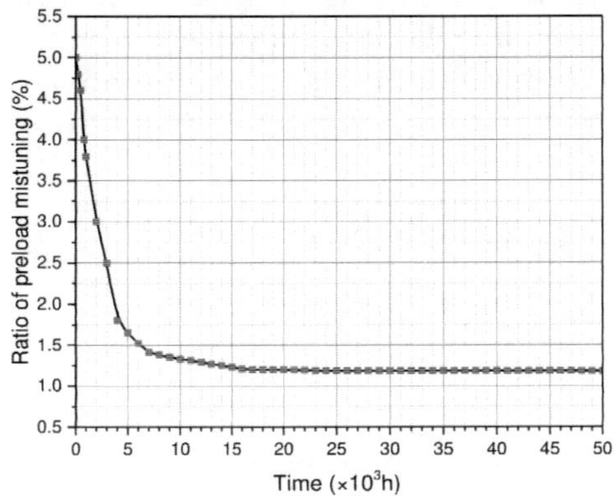

Figure 5 Mistuning ratio versus time under the condition of initial mistuning ratio of 5%

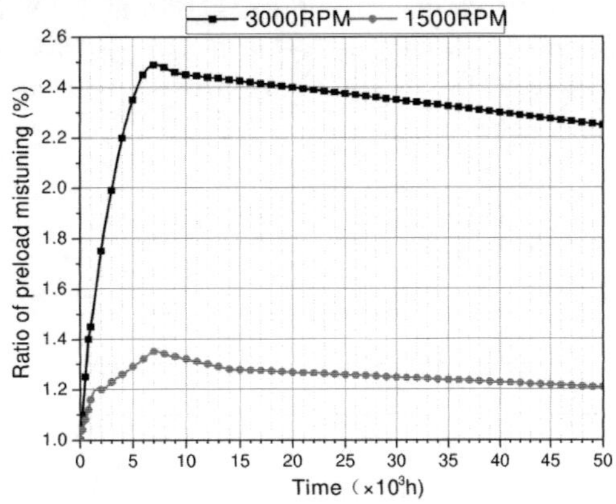

Figure 6. The mistuning ratio versus time under different rotating speeds

3.2. The Mistuning Ratio Under Different Initial Different Rotating Speeds and Temperature

The combined rotor has different working conditions, such as speed, temperature, etc., which affect the connection stiffness caused by the unevenness of the creep relaxation of the rod bolt set. The simulation is carried out for two speed conditions, 3000 rpm and 1500 rpm, respectively. In the simulation process, The ratio of preload mistuning of rod bolt set is changed, as shown in Figure 6. When the speed is large, the mistuning ratio becomes larger, and the time required to reach the threshold is relatively short.

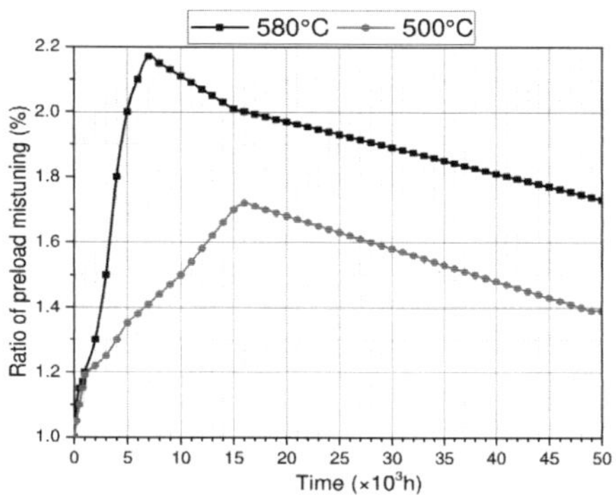

Figure 7. The mistuning ratio versus different temperature

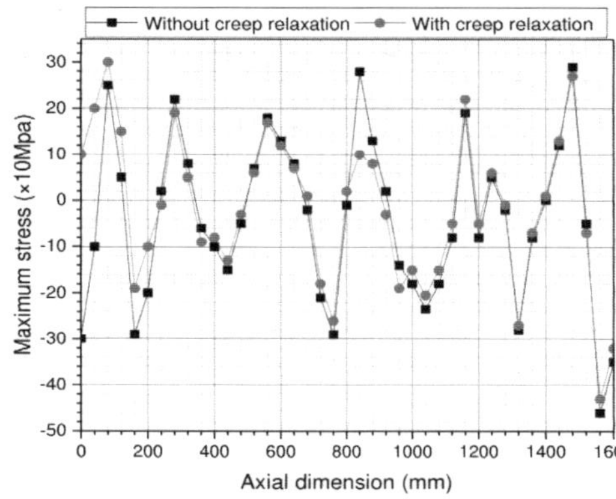

Figure 8. contact stress versus axial time under dimension

Under the condition of initial detuning 1%, the combined rotors with operating temperature of 500 ℃ and 580 ℃ are simulated respectively, as shown in Figure 7. The higher the temperature is, the higher the mistuning ratio is, and the relatively short time it takes to reach the threshold.

3.3. Effect of Creep Relaxation on Structural Strength

For the combined rotor, the maximum contact stress is on the rod bolts, so it is necessary to study the contact stress between the rod bolts and the disk holes. In contrast to no creep relaxation, the variation of contact stress considering creep relaxation is different, as shown in Figure 8. The contact stress on the contact surface between the rod bolt and the disk hole is uneven in the axial direction. Under the influence of the centrifugal force generated by the high-speed rotation, The contact stress is a positive and negative change.

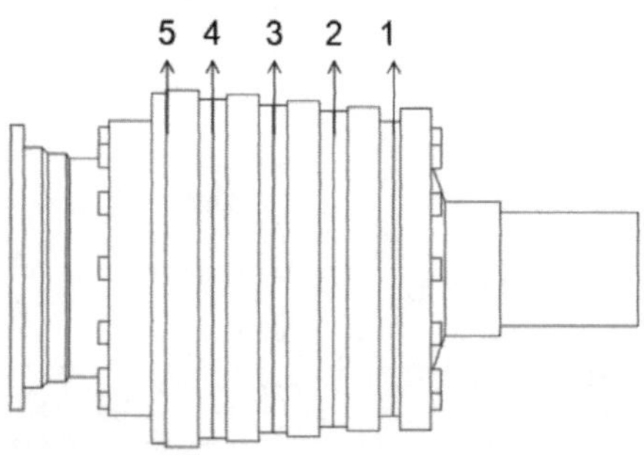

Figure 9. Turbine section model of combined rotor

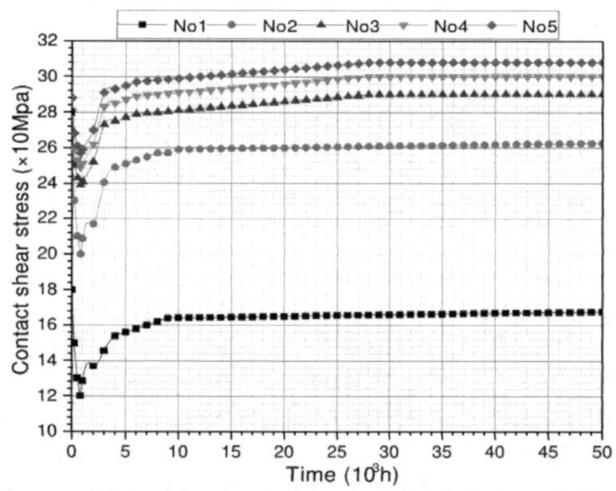

Figure 10. shear stress curve of disks contact surfaces versus time

Due to the lack of preload, the structural strength of the contact surface of the discs is reduced and micro slip occurs. Therefore, it is necessary to study the contact shear stress of disks contact surfaces， as shown in Figure 9 and Figure 10.

The contact stress between the discs of the combined rotor turbine has a transient decrease, and then increases gradually. After the threshold of mistuning ratio, the mistuning ratio decreases gradually and tends to converge.

4. Summary

From the above analysis, we can draw the following conclusions.

1. Under the influence of elastic interaction, the loss of pretightening force of rod bolts at different positions is non-uniform, which will cause the connection of the combined rotor to be out of tune.

2. The mistuning ratio first diverges and then converges, and there is a threshold for the mistuning curve of rod bolts.

3. When the speed is large, the mistuning ratio of rod bolts becomes larger, and the time required to reach the threshold is relatively short. At different temperatures, there is the same trend. As the preload decreases, the structural strength of the combined rotor decreases.

References

[1] S Gao and A Wang 2014 Analysis of Sensitivity of Combined Rotor's Degradation to Crack Position *Noise & Vibration Control* vol 5 pp 156-160

[2] X Li and A Wang 2014 Research on performance degradation of combined rotor considering the fatigue crack in the rod *Chinese Journal of Engineering Design* vol 4 pp382-388

[3] X Li and A Wang 2015 Research on performance degradation of combined rotor considering the microcrack on the rod caused by frequent start-up and stop-down. *Chinese Journal of Engineering Design* vol 2 pp 129-136

[4] H Jingfei and W Chengsheng 2017 Study on Slip of the Screw Thread Subjected to Axial Eccentric Alternate Load *Mechanical Science and Technology for Aerospace Engineering* vol 4 pp 498-503

[5] Y Su and A Wang 2015 Performance degradation of mistuned combined rotor considering rod relaxation *Journal of Vibration and Shock* vol 2 pp 21-26

[6] J He and T Tang 2017 Thermal cycling creep research of combined rotor turbine disc under alternating temperature field *Computer Simulation* vol 1 pp 205-210

[7] Y Sun and C Pang 2014 Research on Rotor Creep Properties under Bending Residual Stress *Journal of Mechanical Engineering* vol 18 pp 162-166

[8] Y Su and A Wang 2014 Effect of rod relaxation upon performance degradation of combined rotor for gas turbine *Chinese Journal of Engineering Design* vol pp 245-250

[9] H Zhang and A Wang 2016 Reseach on performance degradation characteristic of combined rotor considering turbine discs creep *Chinese Journal of Engineering Design* vol 3 pp 417-423

[10] H Liu and W Wang 2010 Influence of creep on low-cycle fatigue strain of Ultra-supercritical steam turbine rotor *Journal of Chinese Society of Power Engineering* vol 9 pp 715-719

[11] Y Zhang and A Wang 2014 Effect of preload mistuning on dynamic characteristic and structural strength of combined rotors *Chinese Journal of Engineering Design* vol 6 pp 534-539

IOP Conf. Series: Materials Science and Engineering **265** (2017) 012025 doi:10.1088/1757-899X/265/1/012025

Parameter identification and optimization of slide guide joint of CNC machine tools

S Zhou and B B Sun*

Mechanical Engineering, Southeast University, Nanjing 211189, China
*Corresponding Author: bbsun@seu.edu.cn

Abstract. The joint surface has an important influence on the performance of CNC machine tools. In order to identify the dynamic parameters of slide guide joint, the parametric finite element model of the joint is established and optimum design method is used based on the finite element simulation and modal test. Then the mode that has the most influence on the dynamics of slip joint is found through harmonic response analysis. Take the frequency of this mode as objective, the sensitivity analysis of the stiffness of each joint surface is carried out using Latin Hypercube Sampling and Monte Carlo Simulation. The result shows that the vertical stiffness of slip joint surface constituted by the bed and the slide plate has the most obvious influence on the structure. Therefore, this stiffness is taken as the optimization variable and the optimal value is obtained through studying the relationship between structural dynamic performance and stiffness. Take the stiffness values before and after optimization into the FEM of machine tool, and it is found that the dynamic performance of the machine tool is improved.

1. Introduction

The machine tool is a complicated structure combined by many parts with different functions and these connection is called joint surface or joint. Studies [1] have shown that joint surface of the machine tool account for 60%-80% of the stiffness, over 90% of the total damping, 40%-60% of the total flexibility and 85%-90% of the static deformation. Therefore, it is an effective way to improve the performance of the machine tool through improving the performance of the joint surface.
Scholars at home and abroad have done extensive research on the parameter identification of joint surface of machine tools. There are mainly three methods to identify parameters, including theoretical calculation method [2], test method [3] and a method combining theoretical calculation and test [4]. However, there are less research on parameter optimization of joint surface and the parameter optimization of joint surface is important to the property of machine tool.
To cure the above problems, this paper adopts optimum design method to identify the joint surface based on the finite element simulation and modal test data firstly. Then sensitivity analysis is used to get the most sensitive parameter of joint surface. At last, optimize this parameter to improve the property of the machine tool.

2. Parameter identification of slide guide joint

2.1. Establishment of parameterized FEM of slide guide joint

A CNC machine tool is taken as the research object in this paper and its 3-D model is shown in Figure 1. Slip joint surface is constituted by the bed and the slide plate. The optimum design module of ANSYS is used to identify the stiffness of slide guide joint surface, and the premise is to establish the parameterized model of slip joint. In practical work, the bed is bolted to the ground through anchor bolt and the quality of the bed is much greater than that of the slide plate. So the bed is simplified into two fixed guide rails and the research object is the slide plate and the slip joint surface constituted by the slide plate and the bed, as shown in Figure 2.

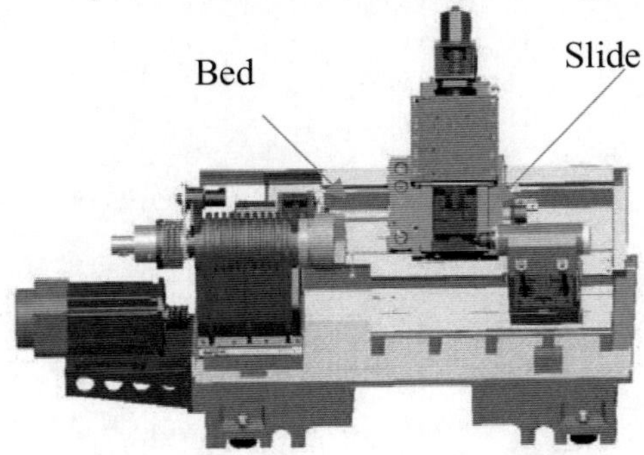

Figure 1. 3D modal of CNC machine tool

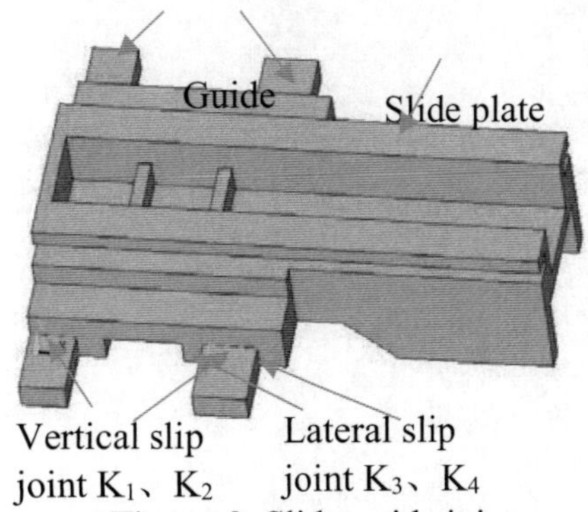

Figure 2. Slide guide joint

The material of slide plate is HT250, the Young modulus is 1.2e5MPa, the density is 7.2e-9t/mm^3 and the Poisson's ratio is 0.25. Solid185 unit is used to mesh the slide plate and the element size is 8. According to Yoshimura Yutakanori, although the contact area of the joint surfaces is different, as long as the average contact pressure is the same, the dynamic data of each unit area of the joint surface can be applied to the general joint surface with the same surface properties. Therefore, the four slip joint surfaces constituted by the bed and the slide plate are respectively equivalent to springs, the stiffness of which are denoted by K_1, K_2, K_3 and K_4, as shown in Figure 2. These springs are meshed with Combine14 and their initial stiffness are all set as 1×10^6N/mm.

2.2. Modal test of slip joint

In order to obtain the low-order natural frequencies and modes of vibration of the slide plate relative to the bed, the bed needs to be bolted to the ground through anchor bolt and the joint surface between the bed and the slide plate should be in accordance with the normal working condition. This experiment adopts the German M+P SO Analyzer data collection and analysis system, and single-input multi-output method is used. Choose the end point of the slide plate as exciting point and multiple response points are arranged evenly on the frame of slide plate. The frequency range of hammer excitation is 0~312.5Hz and the test site is shown in Figure 3. The imaginary frequency curves of driving point in the three directions X, Y and Z obtained in the test are shown in figure 4. The peaks of the imaginary frequency curves correspond to the natural frequencies, which are listed in Table 1.

2.3. Parameter identification of slip joint

Optimization design module of ANSYS has two kinds of optimization method, namely zero order method and first order method [5]. Zero method is a perfect method to deal with most engineering problems effectively. First order method is based on the sensitivity of the objective function to the design variables and is more suitable for accurate optimization analysis. In this paper, zero order method is used for rough optimization in the global design space firstly, and then first order method is used to optimize in the local design space until the most satisfactory solution is found.

In order to identify the stiffness of slip joint, the mathematical model of optimal design is established as follows:

Design variable: $X=\{K_1 \quad K_2 \quad K_3 \quad K_4\}^T$ (1)

K_1 and K_2 stand for the vertical stiffness of slide guide joint surface, and K_3 and K_4 stand for the lateral stiffness of slide guide joint surface.

State variable: $f_{itest}-a \le f_{ical} \le f_{itest}+a$ (2)

Figure 3. Modal test of slide guide joint

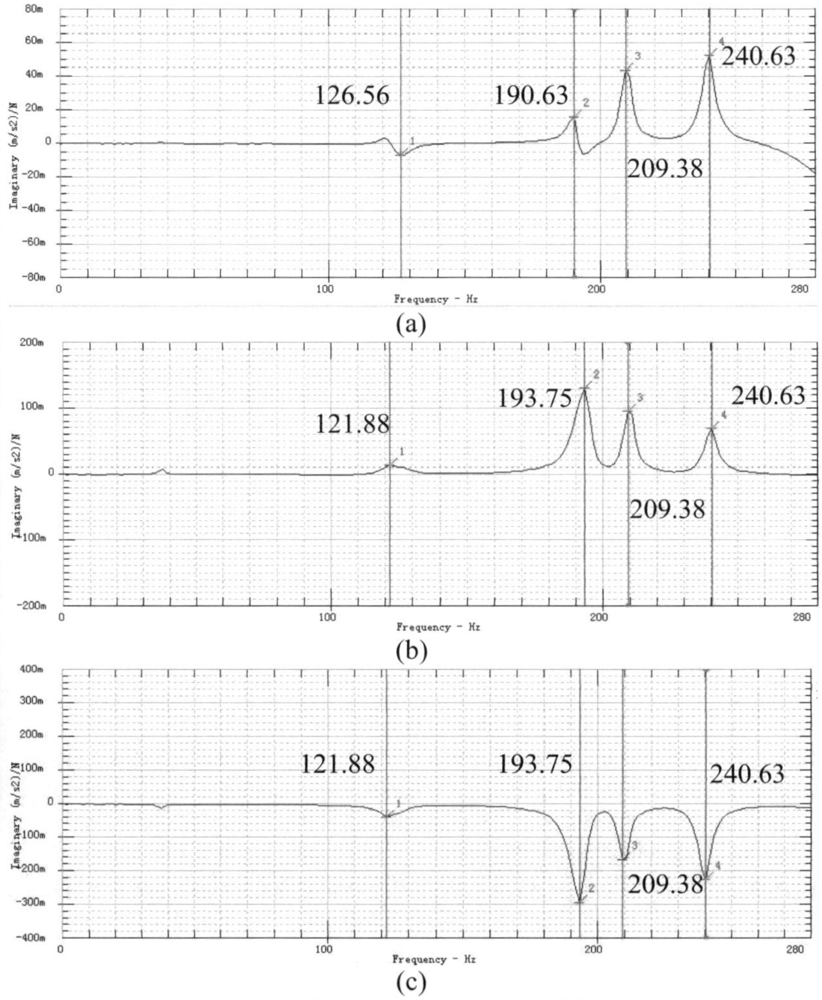

Figure 4. Virtual frequency curves of driving point
(a) X direction (b) Y direction (c) Z direction

a is a constant, f_{itest} is the natural frequencies of joint obtained by modal test, f_{ical} is the natural frequencies of joint obtained by FEA.

Objective function: $$\min F(X)=\sum_{i=1}^{n} a_i\{(f_{ical}-f_{itest})/f_{itest}\}^2 \quad (3)$$

a_i is the weight of the natural frequency of each order and it is chosen according to the contribution of the natural frequency of each order to the objective function.

The first four natural frequencies obtained by the modal test are substituted into the objective function. After a few calculations of iteration in ANSYS, the stiffness values of each slip joint surface is acquired, respectively as:

$$K_1=8.2\times10^5 N/mm,\ K_2=2.0\times10^6 N/mm,\ K_3=8.3\times10^6 N/mm,\ K_4=4.2\times10^5 N/mm.$$

The stiffness values are substituted into the FEM of slide guide joint, and the first four natural frequencies and modes of vibration are gotten. From Table 1, it can be seen that the error of the natural frequency of various orders obtained by FEA and modal test is within 10%, which can be acceptable. From Figure 5, it can be seen that the first order mode of vibration of the slide plate relative to the bed is turning, the second order mode of vibration is yawing, the third order mode of vibration is the coupling of the up and down translation and the left and right translation, and the fourth order mode of vibration is turning. The first four modes of vibration obtained by FEA and modal test are almost consistent, so it can be concluded that the stiffness values of joint surface identified by the method used in this paper is accurate.

Table 1. The comparison between the first four natural frequencies of finite element analysis and modal test of slide guide joint

steps	1	2	3	4
FEA (Hz)	109.53	186.92	215.63	255.20
modal test (Hz)	121.88	193.75	210.94	240.63
errors	10.13%	3.53%	2.22%	6.05%

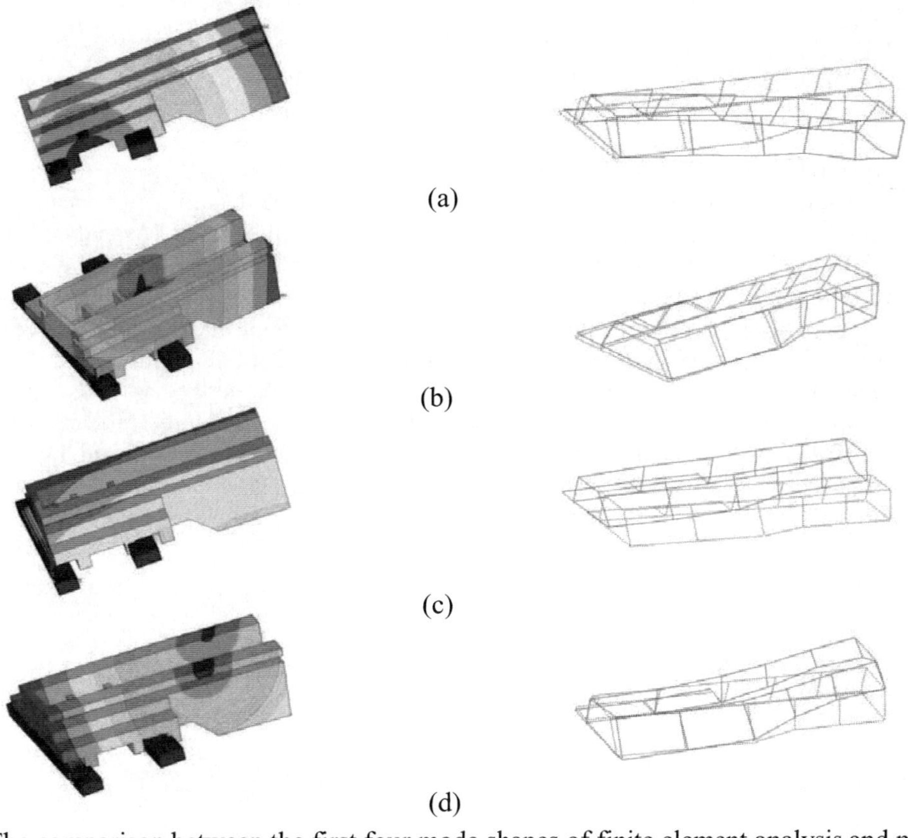

Figure 5. The comparison between the first four mode shapes of finite element analysis and modal test of slide guide joint (a) the first order (b) the second order (c) the third order (d) the fourth order

3. Sensitivity analysis and optimization of stiffness of slip joint surface

3.1. Harmonic analysis of slip joint

In order to find the mode that has the greatest influence on the dynamic performance of the structure, the harmonic analysis of slip joint is carried out. A point on the connecting part of the slide plate and the work table is selected as the observation point, and its displacement response amplitude curve is shown in Figure 6. In the direction X and Z, the maximum peaks appear at the first order natural frequency. In the Y direction, the maximum peak appears in the third order natural frequency. Therefore, the first and the third order modes are the most important modes affecting the dynamic performance of the slip joint.

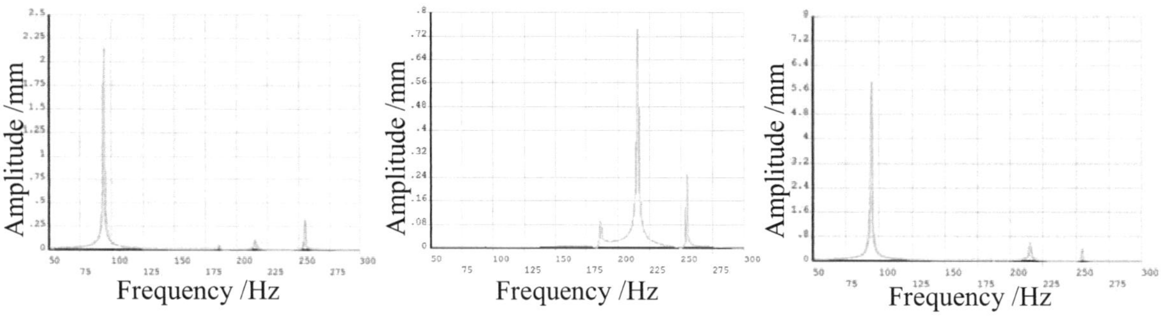

(a) (b) (c)

Figure 6. Displacement response amplitude curve in each direction of connection point on slide carriage (a) X direction (b) Y direction (c) Z direction

3.2. Sensitivity analysis of stiffness of slip joint surface

Since the first and the third order modes have great influence on the dynamic performance of the slip joint, the natural frequency of these two modes are taken as the index to analyse the dynamic sensitivity of stiffness of each joint surface. The design variables are K_1, K_2, K_3 and K_4, the values of which have many uncertainties and randomness for the reason that the stiffness of slip joint is influenced by physical factors such as surface roughness and lubricant medium. Therefore, the design variables follow the Gaussian distribution. Use Latin Hypercube Sampling and Monte Carlo Simulation to simulate, and after sampling 200 times, the natural frequency has been convergent. The sensitivity of dynamic performance of the slide guide joint is shown in Figure 7. It can be seen that the stiffness K_2 has the most impact on the first order natural frequency, followed by stiffness K_1. Likewise, the stiffness K_2 has the most impact on the third order natural frequency, followed by stiffness K_1. In addition, the values of stiffness are all positively correlated with the natural frequencies.

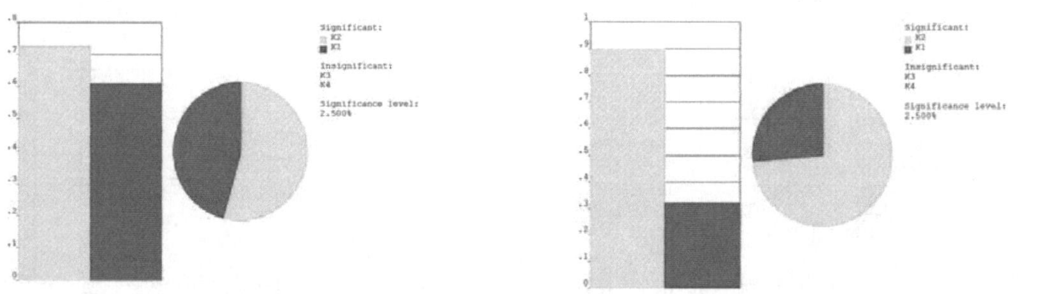

(a) (b)

Figure 7. Sensitivity of dynamic performance of slide guide joint (a) Sensitivity of the first natural frequency (b) Sensitivity of the third natural frequency

3.3. Optimization of stiffness of the slip joint surface

From the sensitivity analysis above, it can be seen that the stiffness K_2 has the most influence on the dynamic performance of slip joint, followed by K_1. Therefore, in order to improve the dynamic performance of slip joint, the values of K_1 and K_2 should be improved. Change separately K_2 and K_1, and observe the changes of the first four steps natural frequency, as shown in Figure 8 and Figure 9. With the increase of K_2, the natural frequency of each order is improved, and the amplitude of improvement decreases gradually. When K_2 equals to 9.5×10^6, the natural frequency of each order tends to be stable. Due to the increase of the stiffness of slip joint will increase the manufacturing cost, 9.5×10^6 can be selected as the optimal value of K_2. Likewise, with the increase of K_1, the natural frequency of each order is improved, and the amplitude of improvement decreases gradually. When K_1

equals to 4×10^6, the natural frequency of each order tends to be stable, so 4×10^6 can be selected as the optimal value of K_1.

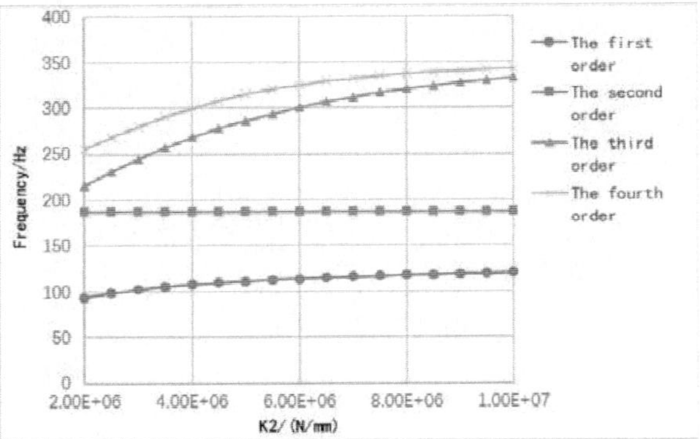

Figure 8. Variation of each natural frequency with stiffness K_2

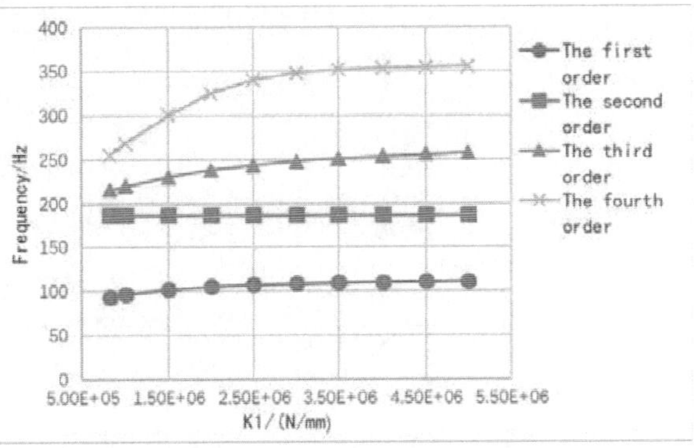

Figure 9. Variation of each natural frequency with stiffness K_1

4. Dynamic analysis of machine tool

A finite element model of machine tool considering joint surface is established. The axial stiffness of ball screw obtained by using the calculation formula in the book named *Basis of ball screw*. Take the stiffness of the slip joint surface before and after optimization into the machine tool, and modal analysis and harmonic analysis are carried out. Table 2 lists the variation of the natural frequencies of the machine tool before and after optimization. Table 3 lists the change of the maximum displacement amplitude of harmonic analysis, and Figure 10 shows the displacement amplitude curves of the machine tool before and after optimization.

It can be seen from Table 2 that the natural frequency of each order is improved in varying degrees after optimization. From Table 3 and Figure 10, it can be seen that the amplitude of displacement response in each direction is decreased to different extent, which shows that the vibration of the machine tool has been restrained.

Table 2. Natural frequencies of machine tool before and after optimization

Steps	1	2	3	4	5	6
Before (Hz)	86.81	89.91	99.22	147.41	149.84	164.87
After (Hz)	87.92	95.58	99.63	149.15	151.72	168.94
Ratio	1.3%	6.3%	0.4%	1.2%	1.3%	2.5%

Table 3. The maximum displacement amplitude of each direction before and after optimization

Direction	X	Y	Z
Before /mm	0.01248	0.00322	0.00854
After /mm	0.01217	0.00263	0.00835
ratio	2.5%	18.3%	2.2%

5. Conclusions

In this paper, the finite element analysis, modal test and optimization method are used to identify and optimize the parameters of slide guide joint surface of CNC machine tool, and the dynamic performance of the machine tool has been improved. The main content and conclusions are as follows.

1) The parametric finite element model of slide guide joint is established firstly, and then optimum design method is adopted to identify the vertical stiffness and lateral stiffness of the slide guide joint surface based on the finite element simulation and modal test data. The accuracy of this method is verified by comparing the results of FEA with that of modal test.

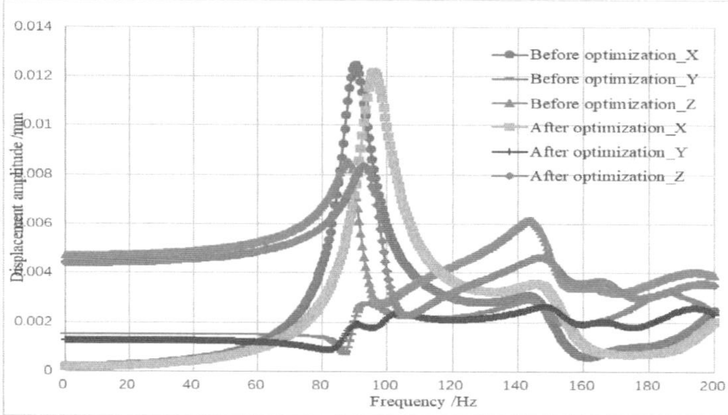

Figure 10. Displacement response amplitude curves before and after optimization

2) From the result of harmonic analysis, the natural frequencies of the first and third order modes have the most influence on the dynamic performance of slip joint. Take these frequencies as objective, the sensitivity analysis of stiffness of each joint surface is carried out using Latin Hypercube Sampling and Monte Carlo Simulation. It is found that the vertical stiffness of slip joint surface constituted by the bed and the slide plate has the most obvious influence on structural dynamic performance.

3) With the increase of K_2 and K_1 respectively, the natural frequency of each order increases and tends to be stable. Find the optimal value of K_2 and K_1 and take the stiffness of the slip joint surface before and after optimization into the machine tool. The results of modal analysis and harmonic analysis show that the dynamic performance of the machine tool has been improved.

Acknowledgments

The work reported in this paper is supported by the Technology Grand Special Project (Approval No. 2013ZX04012032, 2012ZX04002032).

References

[1] Xiao Y C, Wu L S, Ma J F and Hong H 2013 Identifying the unit area dynamic characteristic parameters of fixed joint based on equivalent single DOF system *Applied Mechanics and Materials* **483** 322-5

[2] Mi L, Yin G F, Sun M N and Wang X H 2012 Effects of preloads on joints on dynamic stiffness of a whole machine tool structure *Journal of Mechanical Science and Technology* **26** 495-508

[3] Mehrpouya M, Graham E and Park S S 2013 FRF based joint dynamics modeling and identification *Mechanical Systems and Signal Processing* **39** 265-279

[4] Lin Y and Chen W 2002 A method of identifying interface characteristic for machine tools design *Journal of Sound and Vibration* **255** 481-7

[5] Xue X W, Yuan H and Li S Q 2012 Study on optimizing design of structural rigidity of super-large span continuous rigid frame bridge using ANSYS *Advanced Materials Research* **446-449** 1243-47

IOP Conf. Series: Materials Science and Engineering **265** (2017) 012026 doi:10.1088/1757-899X/265/1/012026

Fatigue Lifespan of Engine Box Influenced by Fan Blade Out

QIU Ju[1], SHI Jingwei[1], SU Huaizhong[1], ZHANG Jinling[1], FENG Juan[1], SHI Qian[1], TIAN Xiaoyu[1]

[1]Beijing Aeronautical Science & Technology Research Institute of COMAC, Future Science and Technology Park, Changping District, Beijing 102211, P. R. China

qiuju@comac.cc

Abstract. This provides precious experience and reliable reference data for future design. This paper introduces the analysis process of Fan-blade-out, and considers the effect of windmill load on the fatigue lifespan of the case. According to Extended Operations (ETOPS) in the airworthiness regulations, the fatigue crack of it is analyzed by the unbalanced rotor load, during FBO. Compared with the lifespan in normal work of the engine, this research provides valuable design experience and reliable reference data for the case design in the near future.

1. Introduction

With the rapid development of modern aviation industry, the safety and reliability of aircraft in particular has caught people's attention. Many painful lessons tell people the aviation history of the world. Once it fails, generally, it influences the performance of the aircraft, but, worst, the machine crashed. As the "heart" of modern aircraft, the aero-engine is prone to all kinds of failures due to its extremely complex structure and its work in harsh environments such as high speed, high temperature and heavy loading [1-2].

In the 1940s, the advent of gas turbine engines had played an important role in the aircraft development from subsonic to supersonic, but 20 years ago, there was a tendency of the design of the engine focused on its performance, reliability and durability, not considering the structure weight, and therefore engine parts were designed to be quite heavy. Especially at the beginning of the 1960s, in order to meet the needs of aircraft speed, high thrust weight ratio, high compression ratio and high temperature requirements of turbine engines, the load of it was greatly improved, and engine vibration was more prominent, resulting in lots of strength, vibration, and fatigue incidents of engine components, such as pylon and cabin. This caused emphasis on the reliability research. Since the 1980s, the structural integrity of aero-engine had been studied in China, and the whole airplane vibration caused by aero-engine was being one of the most critical research projects [3-6]. As for early aero-engine, since the rotor speed was low, the casing stiffness relative to the rotor was relatively high. The coupling vibration of them could be ignored. If only the critical speed of rotor was calculated, this made the casing structure design conservative and heavier; significant characteristics of modern aero engine are high speed, high performance, especially for small and medium sized aircraft engine rotors, which are flexible. The rotor speed may be above the critical speed of the first order (or several orders), the casing wall is thinning, and most of the structures are revolving shells. Dynamic effects between the rotor and the case are very close, mutual coupling and interaction is increasingly strengthening, forming the structure of complex dynamic behaviours, and therefore the whole machine vibration problem is very prominent, so it is necessary to analyze the dynamic characteristics of the engine from the overall angle. In extreme cases, the unbalanced load, due to fan blade out, leads to the fatigue crack of the case. The case vibration accidents quite often occur in military and civil engines to some different degree, resulting in a large number of engines being sent back to the factory for repairs, reducing the service life, and greatly increasing the cost of maintenance. Therefore, it is vital to study the vibration of aero-engine rotor system and whole engine system for the safe and reliable operation of the aero-engine [7].

In the airworthiness, it is put forward that the British defence standard "air gas disaster engine general specification", which detailed rules for blade requirements (blade containment) in the first 12.8.7 and 19.3.5 section. The case must have sufficient strength, working at the highest speed and containing any broken or loose blade. The machine box without large rupture and serious distortion, and the size, quality, speed and trajectory of the escape debris from the engine do not harm the rest of the aircraft. The airworthiness requirements of this point can be learned that the case should be in a safe working range, without major cracks and deformation.

According to CCAR-121 R4 "public air transport carrier of large aircraft of operation qualification examination and approval rules", the definition of (ETOPS), the process is that, on the basis of single expires, the plane ferry flights can be defined as 120 minutes, that is to say, the aircraft case in this phase has no larger crack and deformation.

2. Introduction of engine case

With the development of high speed and high thrust ratio of aero-engine, most of the cases are designed to be thin wall cylindrical structures with weak rigidity. The function of it is to properly support the rotor system according to the designed transmission line, and form a specific airflow passage. For aircraft engines, the gearbox and the rotor system are connected by a number of main bearings, and the case of the rigid rotor will influence the characteristics of the critical speed. In the aviation engine dynamic characteristics analysis, the dynamic performance of the case structure should be considered. At the design stage of the overall engine, the rotor critical speed must be estimated and analyzed so as to analyze the rationality of the overall structure; in engine engineering design stage, the rotor, bearing, case system must also be used to analyse structural dynamic characteristics, in order to determine the accurate range of the critical speed. If Fan Blade Out (FBO) happens, it must ensure that the rotor support system runs normally.

As the composite material is widely used in aircraft, this provides a new choice for case design, from single metal material, gradually developing into a large proportion of composite engine, nowadays. According to the composition of the materials used, the case can be divided into the following types: 1) Aluminum/titanium alloy case; 2) The fiber winding intensifier case; 3) Full composite material case; 4) Ceramic matrix composite material case; 5) Other material case. Although there are various kinds of composite materials, the basic objective is to reduce the weight of the case, improve the strength and stiffness of it, increase engine weight ratio, and strengthen protection effect, etc. In this paper, the thin-walled material of the case is aluminum alloy, and the fatigue performance of the engine system is mainly considered.

3. Introduction of FBO process

3.1. Introduction analysis principle

FBO events are a multidisciplinary and integrated analysis of implicit explicit - implicit process, using MSC. Nastran for FBO effective simulation [8]. FBO is highly nonlinear, consisting of three separate steps. The first step is implicit nonlinear analysis of FBO under the prestressed thrust load (see equation (1)), by using the fine mesh finite element model (FEM), whose result is used as the second step, explicit nonlinear, which is the initial conditions of transient dynamics (see equation (2)). The third step, implicit nonlinear analysis, is engine speed dropping process and the transient state (see equation (3)), by use of the coarse mesh analysis of rotor dynamic characteristics and unbalanced load.

$$KX = F \tag{1}$$

$$Mv = Ft \tag{2}$$

$$M\ddot{X} + C\dot{X} + KX = F(t) \tag{3}$$

Among them, in formula one, K is the structure stiffness matrix; F is the prestressed load. In formula two, M is the structure mass; V is the impact velocity; F is the impact load; T is the impact time. In formula three, M, C and K are structural mass, damping and stiffness matrix, respectively. X is the displacement vector; F (t) is a force varying over time.

3.2.Model of FBO

The thickness of the case is listed in Fig. 1. The transient impact model (see Fig. 2) of FBO includes an engine pod, an engine case, a rotor, blades of the engine, and the rotor axis of the engine providing rotor rotation with three bearings. After FBO, the model (see Fig. 3) of the unbalanced analysis contains nacelle, pylon, and case.

Figure 1. Thickness distribution of engine case.

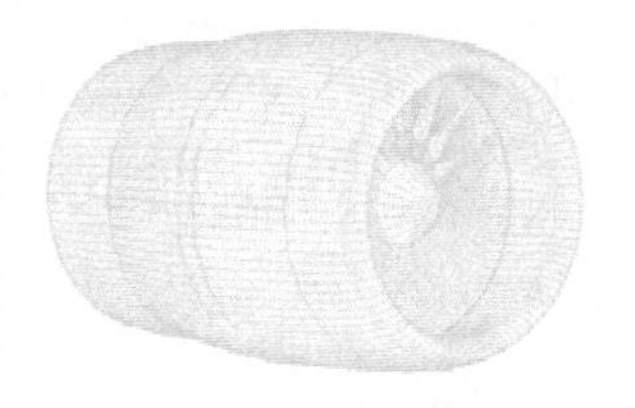

Figure 2. Model of FBO.

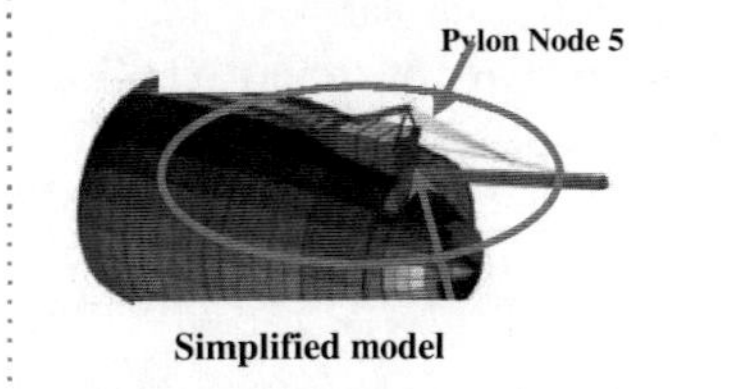

Figure 3. Model of engine rotor dynamics

3.3. Results of FBO

In this paper, two kinds of calculation analysis are carried out, and one is the normal operation of the engine (CASE 1). The second is the FBO model (CASE 2). Figures 4 and 6 show the deformation of the entire engine when the engine turns normally and abnormally, respectively. Figures 5 and 7 are varying load over time in case 1 and case 2.

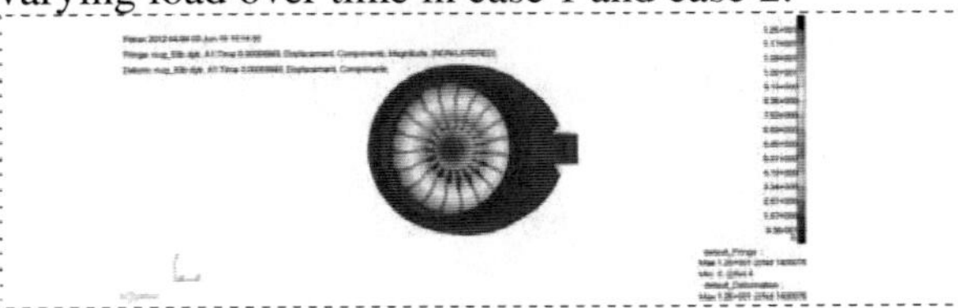

Figure 4. Deformation contour of working engine systems (case 1).

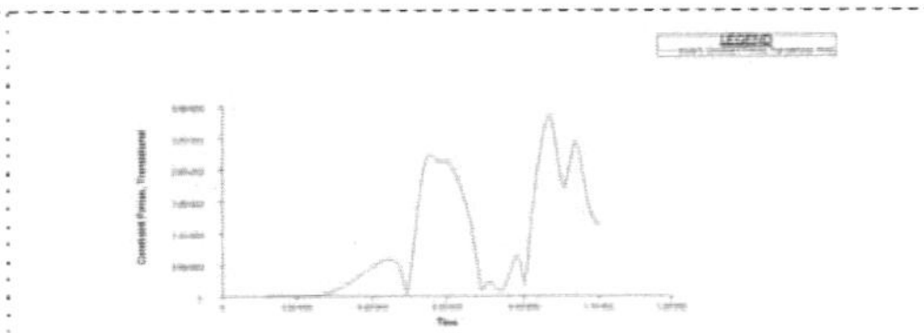

Figure 5. Load at hanged Point changing with time (case 1).

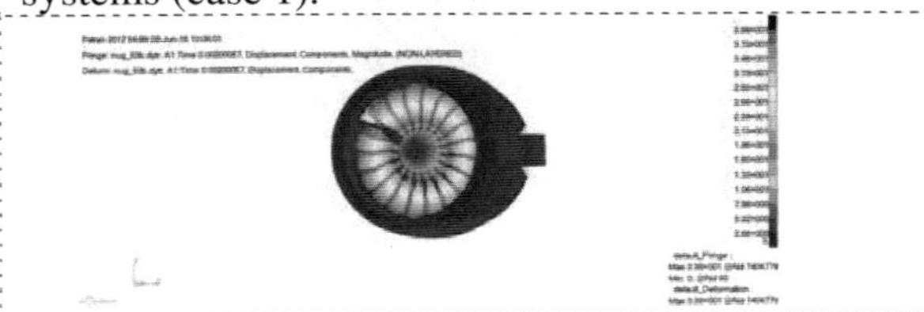

Figure 6. Deformation contour of working engine systems (case 2).

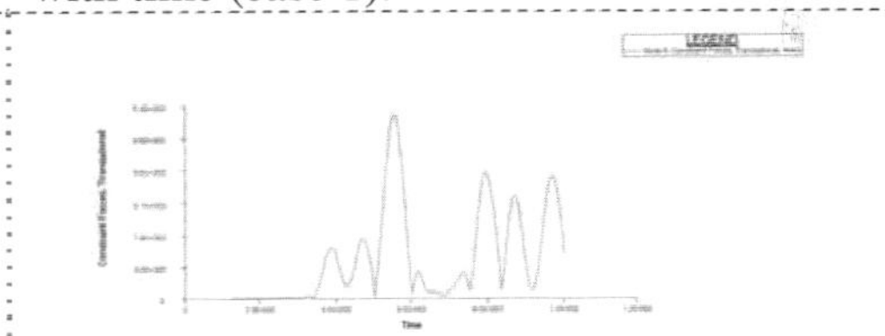

Figure 7. Load at hanged point changing with time (case 2).

Through the above displacement cloud graph and the load variation, the table 1 is summarized for reference.

Table 1 Comparison of CASE 1 and CASE2

Analysis results	Case 1	Case 2
Load at Node 5 (Pound)	0-3900.	0-5400.
Fluctuating numbers of load	2	4
The combined deformation of radial and circumferential direction of engine blades (in)	12.5	39.9

4. Lifespan estimation of engine case

Although the FBO event on commercial jet engines is rare, it must be considered during the design phase of the engine. The engine structure must be able to withstand the large dynamic loads that occur during such extreme loading. As an airworthiness certification process, FAA requires engine manufacturers must successfully verify through practical tests, the engine can withstand an FBO events still located in the installation frame, and the case can continue to work safely.

The case is a typically thin-walled circular shell structure. As the main bearing component of the aero-engine, the shell has the advantages of light weight and large carrying capacity. When the case is impacted by FBO, cracks may appear, which will destroy the integrity of the case, and cause the change of dynamics characteristics. So the fatigue and crack of the case have received extensive attention by scholars.

At present, as for the fatigue life estimation method commonly used in engineering, the S-N curve is widely applied from the stress fatigue. From the strain fatigue, the Manson-Coffin formula is mainly used. After FBO, the stress of the case exceeds the yield limit of the material under the unbalanced rotor vibration, in the (ETOPS) period, and therefore, the strain fatigue life estimation method is adopted.

According to Dirlik equation of the vibration fatigue:

$$E[D]=\sum_i \frac{n_i}{N(S_i)}=\frac{S_t}{k}\int S^b p(S)dS=\frac{E[P]T}{k}\int S^b\left[\frac{S}{4m_0}e^{\frac{-S^2}{8m_0}}\right]dS \quad (4)$$

Among them, N (Si) is the number of stress cycles at time T; ni is the actual counting cycle; Si is the total number of cycles; E [P] is the number of peaks per second;

$$N(S)=E[P]Tp(S) \quad (5)$$

$$p(S)=\frac{\frac{D_1}{Q}e^{\frac{-Z}{Q}}+\frac{D_2 Z}{R^2}e^{\frac{-Z^2}{2R^2}}+D_3 Z e^{\frac{-Z^2}{2}}}{2\sqrt{m_0}} \quad (6)$$

$$D_1=\frac{2(x_m-\gamma^2)}{1+\gamma^2},\quad D_2=\frac{1-\gamma-D_1+D_1^2}{1-R},\quad D_3=1-D_1-D_2,\quad \gamma=\frac{m_2}{\sqrt{m_0 m_4}}$$

$$Q=\frac{1.25(\gamma-D_3-D_2 R)}{D_1},\quad R=\frac{\gamma-x_m-D_1^2}{1-\gamma-D_1+D_1^2},\quad x_m=\frac{m_1}{m_0}\sqrt{\frac{m_2}{m_4}},\quad Z=\frac{S}{2\sqrt{m_0}}$$

Vibration fatigue simulation can be completed by using the finite element analysis in time domain, and get dynamic stress or the strain results.

Through fatigue analysis, the case lifetime during normal operation is shown in the following figure 8:

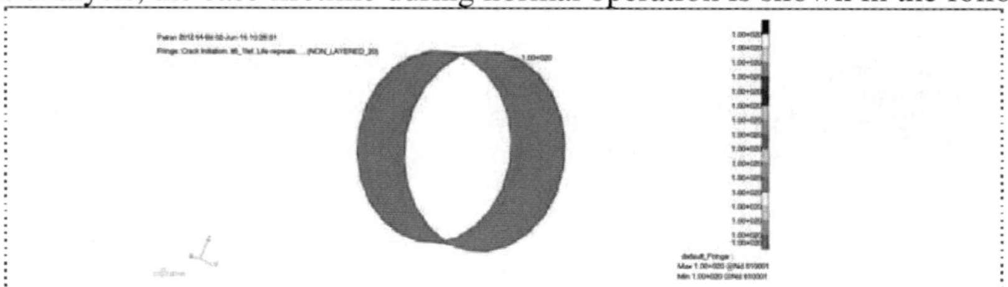

Figure 8. Fatigue lifespan of the case wall under normal work.

The fatigue lifespan and initial crack of the case after FBO are shown below in Figs. 9 and 10.

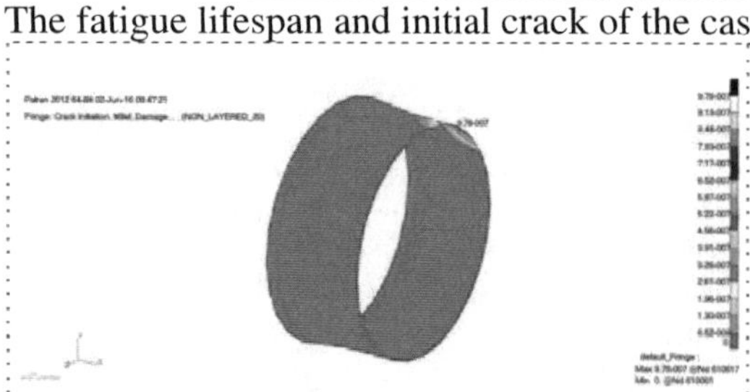

Figure 9. Crack of the case wall after damage

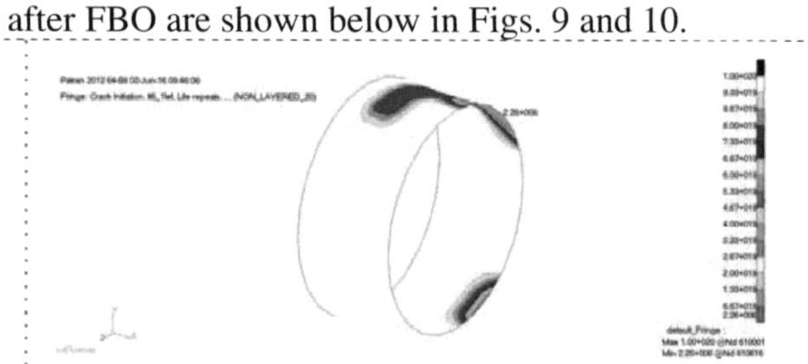

Figure 10. Fatigue lifespan of the case wall after damage

5. Discussion and conclusion

The following conclusions can be obtained from the compared simulation analysis of the CASE 1 and CASE 2.

(1) during FBO, the engine products a large amplitude of fluctuating for the nacelle, which is about 1.5 times as much as that of normal work.

(2) during FBO, the engine causes the numbers of load fluctuation, which is about twice as many as these in the normal working time.

(3) the combined deformation of the radial and circumferential direction of the blade during normal working is 12.5 inches, while that after FBO is 39.9 inches.

(4) when the engine is working normally, the case has no damage and unlimited lifespan; after FBO, the skin of the case is damaged and its life is greatly reduced, but it can still meet the airworthiness requirement.

Acknowledgments

Authors wishing to acknowledge assistance from Aircraft-engine Group in Beijing Aeronautical Science & Technology Research Institute of COMAC, special work by technical staff or financial support of the paper.

References

[1] Wen Dengzhe, Several Dynamic Characteristics of Aero-engine Case Model[D]. Harbin Institute of Technology, 2015, 1-11

[2] Hong J., Chen M., Liu S.G. Application of Whole Engine Finite Element Models in Aero-engine Rotor Dynamic Simulation Analysis[R]. 2007, ASME Paper GT2007227162.

[3] Mathias L, Christophe P, Patrice C. Two-Dimensional Modeling of an Aircraft Engine Structural Bladed Disk-Case Modal Interaction [J]. Journal of Sound and Vibration, 2009, 319(1): 366-391.

[4] LIU Lu-lu , XUAN Hai-jun , ZHANG Na, Investigation on Blade Containment of Aero-engine Composite Case, ENGINEERING MECHANICS, Vol.30 Suppl, June 2013, 314-318. doi: 10.6052/j.issn.1000-4750.2012.06.S093

[5] Zhang G.F., Xu W.N., Xu B., et al. Analytical Study of Nonlinear Synchronous Full Annular Rub Motion of Flexible Rotor-Stator System and its Dynamic Stability [J]. Nonlinear Dynamics, 2009, 57(4): 579-592.

[6] Jiang J, Ulbrich H. Stability Analysis of Full Annular Rub in Rotor-to-Stator Systems [J]. Proceedings of the Applied Mathematics and Mechanics, 2003, 2(1): 88-89.

[7] Ma Shuangchao, Structure Dynamic Analysis and Model Validation of Aero-engine Casings [D], Nanjing University of Aeronautics and Astronautics, December, 2012, 1-6

[8] Mohammad Heidari and David L. Carlson, etc. An Efficient Multi-Disciplinary Simulation of Engine Fan-Blade out Event using MD Nastran, AIAA 2008-2333, 49th AIAA/ASME/ASCE/AHS/ASC Structures, Structural Dynamics, and Materials Conference, 7 - 10 April 2008: 1-12

IOP Conf. Series: Materials Science and Engineering **265** (2017) 012027 doi:10.1088/1757-899X/265/1/012027

Calibration Device Designed for proof ring used in SCC Experiment

X Y Hu, Z Y Kang and Y L Yu

Shanghai Institute of Measurement and Testing Technology, Shanghai, China

Abstract. In this paper, a calibration device for proof ring used in SCC (Stress Corrosion Cracking) experiment was designed. A compact size loading device was developed to replace traditional force standard machine or a long screw nut. The deformation of the proof ring was measured by a CCD (Charge-Coupled Device) during the calibration instead of digital caliper or a dial gauge. The calibration device was verified at laboratory that the precision of force loading is ±0.1% and the precision of deformation measurement is ±0.002mm.

1. Introduction

The stress corrosion crack of material is common in the field of chemical industry, power station, petroleum and ocean exploitation [1-3]. It causes abnormal operation of equipment and large economic loss every year. SCC experiment is a vital method to test the performance of material resistant to SCC. Among the experiment methods recommended by NACE (National Association of Corrosion Engineers) standard TM0177-96, sustained-load method achieved by a proof ring is most widely used for its low cost. The proof ring is used as a loading device, and the load is controlled by measuring its deformation. The new proof ring has a good linearity, and the force load can be measured according to the factory inspection report. After a long-term service, proof ring always has a nonlinearity. The relationship between deformation and force should be verified by fixed interval calibration [4].

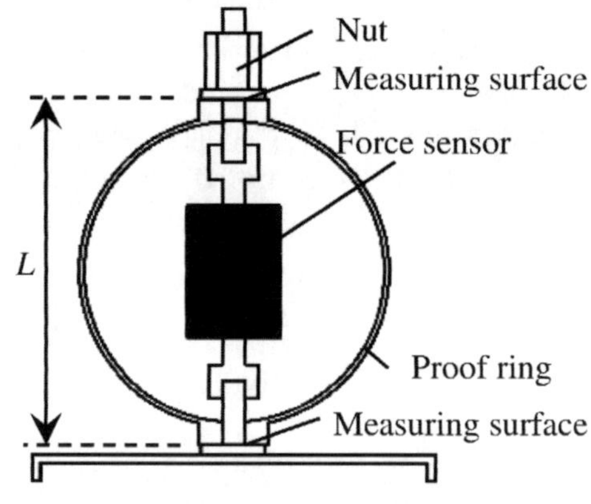

Figure 1. Traditional calibration device

The traditional calibration device is shown in figure 1. The proof ring is deformed by a nut. The force is detected by a force sensor and the deformation $\triangle L$ is measured manually by a digital caliper. The traditional method is always unsatisfied [5-6]. The first reason is the proof ring is not made from anti-corrosion material. The measuring surface is easily corroded, increasing the error. The second is that the deformation is measured manually, so the manual error is inevitable and working efficiency is low. To Therefore, an automatic calibration device for proof ring is proposed in this paper. The structure and working principle is depicted and its performance is verified.

2. Design of loading device

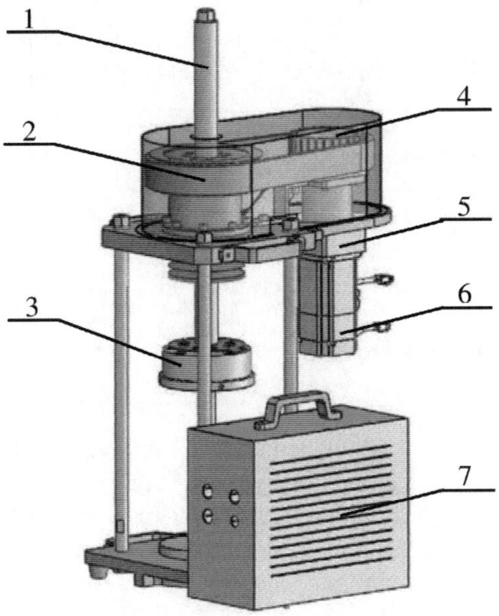

1. Trapezoidal screw 2. Driven wheel 3. Force sensor 4. Driving wheel 5. Planetary reducer 6. Servo 7. Electrical control box

Figure 2. Structure of the loading device

The schematic map of the loading device is shown in figure 2. It is composed of trapezoidal screw, driven wheel, force sensor, driving wheel, planetary reducer, servo and electrical control box. Two wheels are driven by servo through synchronous belt, the advantages of which are low noise and high efficiency. The trapezoidal screw moves along with the driven wheel to realize the force loading, which is measured by the force sensor amounted on the end of the screw. A planetary reducer is applied to achieve precision loading. Two pressure plates are made of heat-treated high carbon alloy tool steel possessing high hardness. The support parts and wheels are made of 2A12 aviation aluminium in order to reduce weight. The dimension of the device is 300*430*800mm and the weight is 46kg. The loading device possesses the merits of compact structure, high efficiency and low weight.

3. Deformation Measurement of the Device

The working principle of the device is shown in figure 3. The calibration plate which has a circular feature symbol is mounted on the upper part of the proof ring by a magnet. The CCD camera is located on a tripod. The mounting height of the CCD camera has no special requirement as long as the feature symbol is in the view of CCD before and after the deformation. The proof ring is loaded by the loading device. When loaded, the underpart of the proof ring is fixed on the pressure plate and the upper part with the calibration plate has a displacement. The position of the feature symbol in the CCD camera changes and the value can be measured by correlating two images before and after loading.

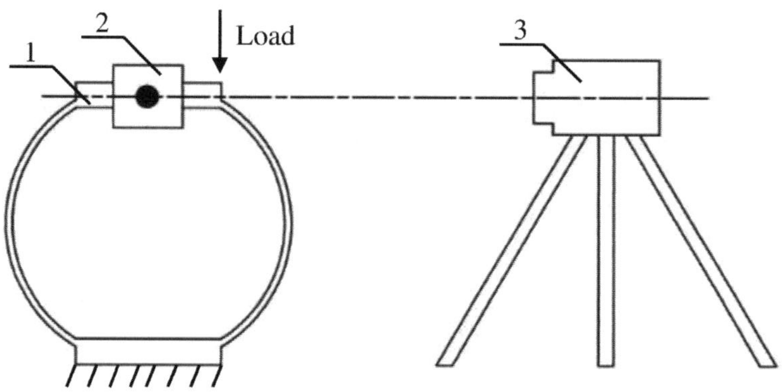

1. Proof ring 2. Calibration plate 3. CCD camera

Figure 3. Deformation Measurement of the Proof Ring

4. Verification of the Calibration Device

4.1 Loading Force

To verify the performance of the loading device, a lever-type force standard machine is employed. The parameters of the force standard machine is listed in table 1. The calibration procedure is in accordance with "JJG 144-2007 Verification Regulation of Standard Dynamometers". The whole loading process is divided into 6 steps. For each step, a 5kN force loading is applied. The result is shown in figure 4 that the max error is less than 0.08%.

Table 1. Parameters of Force Standard Machine

Model	Measurement Range	Max Permitted Error
EE1-10	(1~100)kN	±0.03%

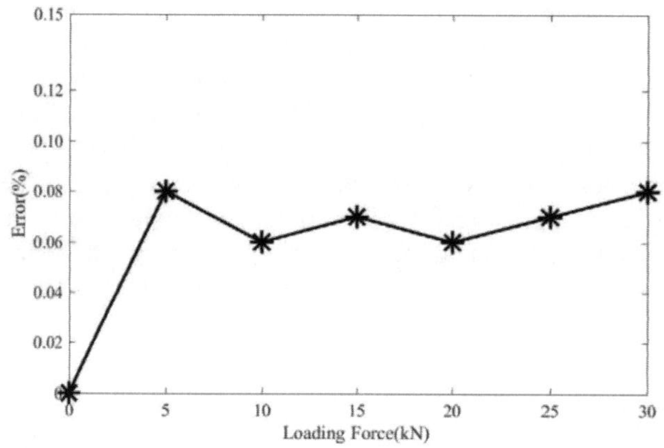

Figure 4. Measurement Error of the Loading Device

4.2 Deformation Measurement

The deformation measurement of the device is verified by a laser interferometer. The parameters of the CCD camera and the laser interferometer is listed in table 2 and table 3. The loading device is located on a desk. The proof ring with the calibration plate is put between the upper and under pressure plates. The CCD camera on the tripod aims at the feature symbol. The interference mirror of the laser interferometer is put on the desk and the reflector is fixed on the force sensor magnetically. The position of the interference mirror is adjusted according to the reflector to make sure that the receiver can detect the distance information. When the proof ring is loaded, the force sensor moves along with the upper pressure plate. The displacement of the force sensor which is equal to the deformation of the proof ring can be measured by the interferometer mirror. Take result of the laser interferometer as benchmark, the error of the designed device is shown in figure 5. The maximum error is 1.9μm.

Table 2. Parameters of CCD Camera

Manufacturer	Resolution	Sensor
DALSA	1600*1200	1/3 inch

Table 3. Parameters of Laser Interferometer

Manufacturer	Model	Measurement Range	Max Permitted Error
Renishaw	XL-80	(0~40)m	$\pm 0.5*10^{-6}L$ *L*:Stroke

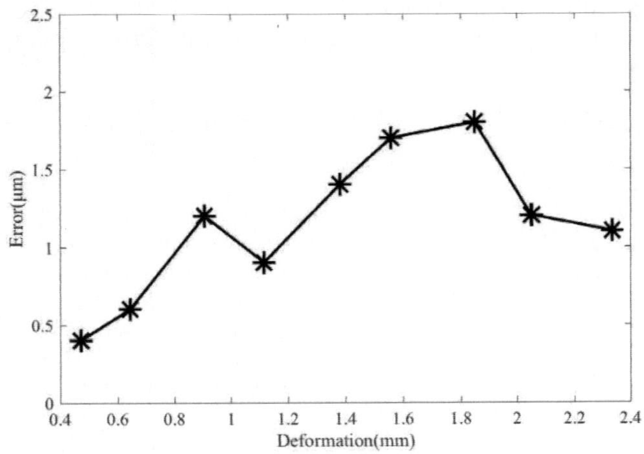

Figure 5. Measurement Error of the Deformation

5. Conclusions:

In this paper, a calibration device for proof ring is designed and verified. A compact size loading device is developed. The CCD is introduced to realize automatic deformation measurement. The designed calibration device is verified by the force standard machine and laser interferometer. The result shows that the maximum error of force measurement is 0.08% and that of deformation measurement is 1.9μm.

References

[1] Han YL, Han EH, Peng QJ and Wei K 2017 *Corros Sci* **121** 1-10
[2] Ma WJ and Soghrati S 2017 *Corros Sci* **125** 87-98
[3] Tan YS, Yu HF, Ma HY, Zhang YQ and Wu CY 2017 *Constr Build Mater* **141** 453-460
[4] Huang ZY, Yuan PB, Tao TJ and Li GB 2010 *Phys Testing Chem Anal Part A* **3** 183-185
[5] Zhang TT, Lu CH 2016 *Bao Tech* **3** *63-66*
[6] Herve M, Nicolas D 2009 *NACE Corrosion 2009* 09094

IOP Conf. Series: Materials Science and Engineering **265** (2017) 012028 doi:10.1088/1757-899X/265/1/012028

Accuracy of Automatic Cephalometric Software on Landmark Identification

Anuwongnukroh N[1], Dechkunakorn S[1*], Damrongsri S[2], Nilwarat C[3], Pudpong N[3], Radomsutthisarn W[3] and Kangern S[3]

[1]Department of Orthodontics, Faculty of Dentistry, Mahidol University, Bangkok, Thailand
[2]Department of Oral and Maxillofacial Radiology, Faculty of Dentistry, Mahidol University, Bangkok, Thailand
[3]Faculty of Dentistry, Mahidol University, 6 Yothee Road, Rajthevee, Bangkok, Thailand

*Corresponding Author: Dechkunakorn S E-mail: surachai.dec@mahidol.ac.th

Abstract. This study was to assess the accuracy of an automatic cephalometric analysis software in the identification of cephalometric landmarks. Thirty randomly selected digital lateral cephalograms of patients undergoing orthodontic treatment were used in this study. Thirteen landmarks (S, N, Or, A-point, U1T, U1A, B-point, Gn, Pog, Me, Go, L1T, and L1A) were identified on the digital image by an automatic cephalometric software and on cephalometric tracing by manual method. Superimposition of printed image and manual tracing was done by registration at the soft tissue profiles. The accuracy of landmarks located by the automatic method was compared with that of the manually identified landmarks by measuring the mean differences of distances of each landmark on the Cartesian plane where X and Y coordination axes passed through the center of ear rod. One-Sample T test was used to evaluate the mean differences. Statistically significant mean differences ($p<0.05$) were found in 5 landmarks (Or, A-point, Me, L1T, and L1A) in horizontal direction and 7 landmarks (Or, A-point, U1T, U1A, B-point, Me, and L1A) in vertical direction. Four landmarks (Or, A-point, Me, and L1A) showed significant ($p<0.05$) mean differences in both horizontal and vertical directions. Small mean differences (<0.5mm) were found for S, N, B-point, Gn, and Pog in horizontal direction and N, Gn, Me, and L1T in vertical direction. Large mean differences were found for A-point (3.0 < 3.5mm) in horizontal direction and L1A (>4mm) in vertical direction. Only 5 of 13 landmarks (38.46%; S, N, Gn, Pog, and Go) showed no significant mean difference between the automatic and manual landmarking methods. It is concluded that if this automatic cephalometric analysis software is used for orthodontic diagnosis, the orthodontist must correct or modify the position of landmarks in order to increase the accuracy of cephalometric analysis.

1. Introduction

Cephalometric radiography is an essential tool in the diagnosis and treatment of malocclusions and skeletal discrepancies. It can be used to diagnose, treatment plan, and evaluate treatment results [1]. Treatment planning decisions are made based on linear and angular measurements called cephalometric analysis using landmarks obtained from lateral skull X-rays. Traditionally, cephalometric analysis has been completed by a manual approach. As radiographs have become progressively digital, computer programs have been developed to aid in cephalometric analysis. More recently, some computer software programs have introduced a fully automatic mode. This mode allows for automatic localization of the landmarks used in these analyses. Automatic identification of landmarks of these programs has been undertaken in different ways and can be classified into four categories: (1) image filtering plus knowledge-based landmark search, (2) model-based approach, (3) soft-computing or learning approach, and (4) hybrid approach [2]. Different effectiveness in landmark localization by automatic mode have been reported in various studies [3-10]. However, accuracy of the dental imaging software (Carestream Dental, version 6.14) which is a fully automatic cephalometric analysis program available at Mahidol University is not yet reported. Therefore, it was the aim of this study to evaluate the accuracy of this software program in locating the cephalometric landmarks.

2. Materials and Methods

The ethics approval for this study was obtained from the Faculty of Dentistry/Faculty of Pharmacy, Mahidol University, Institutional Review Board. Thirty lateral cephalograms of orthodontic patients were randomly selected from the database of the Oral and and Maxillofacial Radiology Clinic, Faculty of Dentistry, Mahidol University.

The criteria for selecting the subjects were as follows: 1) cephalometric radiographs taken with the same x-ray machine (CS9000C); 2) high quality radiographs without artifacts; 3) lateral cephalograms with fully intact permanent central incisors and no craniofacial deformities, such as cleft lip and cleft palate, etc.

Manual cephalometric method was used as a gold standard of which the measurements were compared with the automatic mode. For the manual tracing, an acetate paper was overlaid on the lateral cephalograms and the outline of skull, facial structures and teeth were traced using a sharpened 0.5 mm HB pencil over a light box by one examiner[g]. Then, 13 anatomical landmarks (Fig. 1) were identified and a consensus were made between two experienced orthodontists [a,b]. The thirteen landmarks (Fig. 1) used in this study were Sella (S), Nasion (N), Orbitale (Or), A- point (A-pt), B-point (B-pt), Pogonion (Pog), Gnathion (Gn), Menton (Me), Gonion (Go), Upper incisal tip (U1T), Upper incisal apex (U1A), Lower incisal tip (L1T), and Lower incisal apex (L1A).

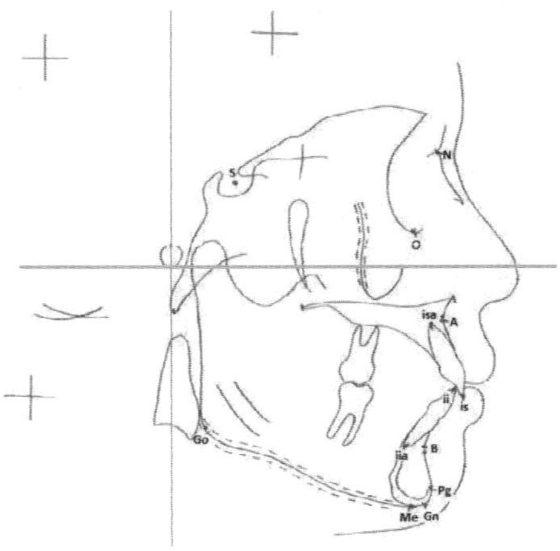

Fig. 1 The thirteen landmarks used in this study.

For the automatic software, all cephalograms were automatically analyzed by the software. Then, the printouts (1:1 ratio) which included tracing lines and landmarks were used for superimposition with the tracings from the manual method. Superimposition of each subject was obtained by registration at the soft tissue profile. The bisecting line from the image of the cephalostat through the center of the machine ear rod was defined as y-axis, and the line perpendicular to the y-axis through the center of the ear rod was defined as x-axis [9]. This coordinate was used for distance measurements of all the 13 landmarks for both manual and automatic methods.

All measurements were done by 2 examiners e,f with the same cephalometric protractor. Mean of the measurements by 2 examiners was calculated and used for statistical analysis.

Mean distances in x- and y-axes of each landmark in the manual tracing were used as a baseline to compare with those obtained from the automatic method. Descriptive statistics was used for each variable and the mean different distances of each landmark in horizontal and vertical direction were evaluated for significant differences by One-Sample T test.

3. Results

The accuracy of the automatic software program in locating the landmarks is described as the distance the automatic software's estimate is from the clinicians' estimate. Mean differences for automatic and manual landmark identification in horizontal (x-axis) and vertical (y-axis) directions are shown in Tables 1 and 2. There were no statistically significant difference ($p<0.05$) between manual and automatic methods in 8 of 13 landmarks (61.54%: S, N, U1T, U1A, B, Gn, Pog, and Go) in horizontal direction, 6 of 13 landmarks (46.15%: S, N, Gn, Pog, Go, and L1T) in vertical direction and 5 of 13 landmarks (38.46%: S, N, Gn, Pog, and Go) in both directions. Statistically significant differences ($p<0.05$) were found in mean differences of 5 of 13 landmarks (38.46%: Or, A-point, Me, L1T, and L1A) in horizontal direction, 7 of 13 landmarks (53.85%: Or, A-point, U1T, U1A, B-point, Me, and L1A) in vertical direction and 4 of 13 landmarks (30.77%: Or, A-point, Me, and LIA) in both directions. The S, N, B-pt, Gn, and Pog in horizontal direction and N, Gn, Me, and L1T in vertical direction were the most accurately identified landmark because the landmark location discrepancy was found to be < 0.5 mm. The A-pt in horizontal direction (location discrepancy of 3.0 < 3.5 mm) and the L1A in vertical direction (location discrepancy of > 4.0 mm) were the least accurately identified landmarks (Table 3).

Table 1. Comparison of the means differences for automatic and manual landmark identification in horizontal direction (x-axis)

Land marks	Horizontal (x-axis)						
	Software		**Manual**		**Differences**		**Sig**
	Mean(mm)	**SD(mm)**	**Mean(mm)**	**SD(mm)**	**Mean(mm)**	**SD(mm)**	
S	19.40	3.44	19.38	3.46	0.02	1.14	0.96
N	82.01	5.75	82.0	5.78	0.01	0.98	0.96
Or	71.0	5.04	72.48	5.43	-1.48	2.55	0.003*
A	87.05	5.46	83.72	5.88	3.33	1.77	<0.001*
U1T	91.53	5.85	92.11	5.69	-0.58	1.71	0.76
U1A	80.16	5.21	79.58	5.92	0.58	2.09	0.14
B	81.27	5.61	81.25	5.77	0.017	0.77	0.91
Gn	79.13	5.67	78.96	6.05	0.18	1.37	0.49
Pog	81.47	5.81	81.5	6.08	-0.03	0.76	0.81
Me	75.33	5.65	73.78	5.95	1.54	1.44	<0.001*
Go	14.41	2.91	13.89	3.48	0.52	1.95	0.16
L1T	87.96	5.49	88.83	5.42	-0.87	1.92	0.02*
L1A	75.6	5.57	78.06	5.69	-2.46	2.13	<0.001*
Total mean difference					**0.89**	**1.58**	

*Significant difference at p< 0.05

Table 2. Comparison of the means differences for automatic and manual landmark identification in vertical direction (y-axis).

Land marks	Vertical (y-axis)						
	Software		**Manual**		**Differences**		**Sig**
	Mean(mm)	**SD(mm)**	**Mean(mm)**	**SD(mm)**	**M(mm)**	**SD(mm)**	
S	27.91	4.06	27.04	4.39	0.87	2.03	0.27
N	35.73	4.98	35.93	5.89	0.19	2.37	0.661
Or	7.67	4.13	10.18	5.07	2.51	2.89	<0.001*
A	20.96	4.58	20.04	4.77	0.92	1.89	0.013*
U1T	42.09	4.99	41.35	4.68	0.74	1.45	0.009*
U1A	19.79	4.43	21.66	4.68	1.87	2.09	<0.001*
B	59.66	5.21	57.48	5.18	2.18	1.90	<0.001*

Gn	74.93	5.51	75.19	5.67	0.26	1.24	0.264
Pog	69.13	5.43	70.24	6.12	1.11	2.26	0.12
Me	76.43	5.54	76.01	5.37	0.43	0.69	0.002*
Go	45.69	3.55	47.0	4.52	1.31	3.75	0.66
L1T	38.61	5.17	38.76	4.94	-0.15	2.03	0.688
L1A	60.55	5.1	55.83	4.92	4.73	1.93	<0.001*
Total mean difference					1.32	2.04	

*Significant difference at $p < 0.05$

Table 3. Landmark location discrepancy between automatic and manual method in horizontal and vertical directions.

Degree of discrepancy (mm)	Landmarks	
	Horizontal	Vertical
$x < 0.5$	S, N, B-pt, Gn, Pog	N, Gn, Me, L1T
$0.5 \le x < 1.0$	U1T, U1A, Go, L1T	S, A-pt, U1T
$1.0 \le x < 1.5$	Or, Me	Pog, Go
$1.5 \le x < 2.0$	-	UIA
$2.0 \le x < 2.5$	L1A	Or, B-pt
$2.5 \le x < 3.0$	-	-
$3.0 \le x < 3.5$	A-pt	-
$3.5 \le x < 4.0$	-	-
$x > 4.0$	-	L1A

4. Discussion

Manual cephalometric analysis is a well established traditional method used as a quality standard and reference [11, 12]. In our study, two experienced orthodontists manually identified landmarks with a consensus in order to avoid professional bias. These landmarks were then used as baseline landmarks and compared with those located by the automatic system.

According to statistical analysis, significant mean differences ($p<0.05$) on the x- and y-axes for each landmark would represent the error of the automatic system in locating the landmarks. The results in this study showed that errors between manual and automatic systems were not significantly different ($p<0.05$) for 5 of 13 landmarks (S, N, Gn, Pog, and Go). This is consistent with the study of Liu et al [9] who also reported that 5 of 13 landmarks (S, N, Po, Or, and Gn) showed no significant difference in landmarking. One of the contributing factors for the insignificant difference may arise from the reason that these landmarks were in clear border of the craniofacial structure. However, if the landmark is in a blurred area of the craniofacial structures, the error will be large [13, 14].

Regarding the magnitude of cephalometric landmarks identification error, Trpkova et al [15] recommended that 0.59 mm of total error for the x-coordinate and 0.56 mm for the y-coordinate are within acceptable levels of accuracy. In our study, total mean differences in x-axis and y-axis were 0.89 and 1.32 mm, respectively. The results suggested that computerized landmarking of this software was not accurate enough to allow its use for clinical purpose. Variation in magnitude of error in automatic landmarking were reported in other previous studies [3-6,9,10]. Factors influencing the magnitude of error could be summarized as: (1) errors associated with the automatic system techniques, (2) quality of the digital images, and (3) superimposition of the nearby anatomic structures.

In this study, some cephalometric points showed less error on the horizontal plane (S, N, B-pt, Gn, and Pog) and some showed better results on the vertical plane (N, Gn, Me, and L1T). This is in accordance with the statement that the distribution of errors for many landmarks is systematic and follows a typical pattern (non-circular envelop). In fact, it has been reported that some cephalometric landmarks are more reliable in the horizontal dimension whereas others are more reliable in the vertical dimension [15]. The results in this study revealed that B-pt could be identified with a higher degree of accuracy in horizontal plane than vertical plane, whereas A-pt showed a higher degree of accuracy in the vertical plane than horizontal plane.

Regarding the pattern of error, landmarks lying on horizontal edges or planes were more accurately located in the vertical dimension. Similarly, landmarks lying on vertical edges were more accurately located in the horizontal dimension. It was also noted that there were large mean errors (discrepancy of > 2 mm.) in some landmarks (U1A, L1A, Or, B-pt, and A-pt). The possible reason may be because those landmarks were located on poorly defined structures or were overlapped with surrounding structures.

More accurate algorithms of automated landmarking are needed for the improvement of the software so that it can be reliably used in orthodontics. The error of automatic landmark identification must be less than or close to that of manual identification.

5. Conclusion

The automatic cephalometric analysis software in this study was unable to compete with manual identification of radiographs in terms of accuracy of landmark identification. In fact, the ability to automatically identify landmarks is fair for many landmarks, but overall, the software was not accurate enough for routine clinical use. The orthodontist must use the fully automatic cephalometric analysis system with caution.

References

[1] W. R. Proffit, D. M. Sarver, J. L. Ackermann, Orthodontic diagnosis: the development of a problem list. Contemporary Orthodontics, Mosby Elsevier, Philadelphia. 2007, pp. 201-211.
[2] R. Leonardi, D. Giordano, F. Maiorana, C. Spampinato. Automatic cephalometric analysis: A systematic review. Angle Orthodont. 78 (2008) 145-151.
[3] S. Parthasarathy, S. T. Nugent, P.G. Gregson, D. F. Fay, Automatic landmarking of cephalograms. Comput. Biomed. Res. 22 (1989) 248-269.
[4] J. Ren, D. Lid, D. Feng, J. Shao. A knowledge-based automatic cephalometric analysis method. Conference proceedings: 20th International Conference of the IEEE Engineering in Medicine and Biology Society. IEEE, 20(2) (1998) 723–727.
[5] T. J. Hutton, S. Cunningham, P. Hammond. An evaluation of active shape models for the automatic identification of cephalometric landmarks. Eur. J. Orthod. 22 (2000) 499–508.
[6] D. J. Rudolph, P. M. Sinclair, J. M. Coggins. Automatic computerized radiographic identification of cephalometric landmarks. Am. J.Orthod. Dentofacial. Orthop. 113 (1998) 173–179.
[7] V. Ciesielski, A. Innes, J. Sabu, J. Mamutil. Genetic programming for landmark detection in cephalometric radiology images. Int. J. Knowl. Based Intell. Eng. Syst. 7 (2003) 164–171.
[8] I. El-Feghi, M. A. Sid-Ahmed, M. Ahmadi. Automatic localization of craniofacial landmarks for assisted cephalometry. Pattern Recognit. 37 (2004) 609–621.
[9] J. Liu, Y. Chen, K. Cheng. Accuracy of computerized automatic identification of cephalometric landmarks. Am. J. Orthod. Dentofacial. Orthop. 118 (2000) 535–540.
[10] V. Grau, M. Alcaniz, M. C. Juan, C. Monserrat, C. Knoll. Automatic localization of cephalometric landmarks. J. Biomed. Inform. 34 (2001) 146–156.
[11] T. Sommer, R. Ciesielski, J. Erbersdobler, W. Orthuber, H. Fischer-Brandies. Precision of cephalometric analysis via fully and semiautomatic evaluation of digital lateral cephalographs. Dentomaxillofac. Radiol. 38 (2009) 401-406.
[12] D. Watcharinporn, S. Dechkunakorn, N. Anuwongnukroh, C. Sinthanayothin, P. Santiwong, W. Bholsithi, et al. Assessment of the Accuracy and Reliability of Cephsmile v.2 in Cephalograms and Model Measurements. LNEE, 143 (2012) 413-422.
[13] S. Baumrind, R. C. Frantz. The reliability of head film measurements. 1. Landmark identification. Am. J. Orthod. 60 (1971) 111-127.
[14] J. Broch, O. Slagsvold, M. Rosler. Error in landmark identification in lateral radiographic headplates. Eur. J. Orthod. 3 (1981) 9-13.
[15] B. Trpkova, P. Major, N. Prasad, B. Nebbe. Cephalometric landmarks identification and reproducibility: A meta analysis. Am. J. Orthod. Dentofacial. Orthop. 112 (1997) 165-170.

IOP Conf. Series: Materials Science and Engineering **265** (2017) 012029 doi:10.1088/1757-899X/265/1/012029

Quantitative Evaluation of Heavy Duty Machine Tools Remanufacturing Based on Modified Catastrophe Progression Method

Li shunhe[1],Rao jianhua[1,3],Gui lin[2],Zhang weimin[1] and Liu degang[1]

[1]China University of Geosciences,Wuhan,Hubei 430074.China
[2]Wuhan Heavy Duty Machine Tool Group Corporation,Wuhan,Hubei 430077.China
[3]Corresponding Author. E-mail address: raojianhcug@163.com

Abstract:The result of remanufacturing evaluation is the basis for judging whether the heavy duty machine tool can remanufacture in the EOL stage of the machine tool lifecycle management.The objectivity and accuracy of evaluation is the key to the evaluation method.In this paper, the catastrophe progression method is introduced into the quantitative evaluation of heavy duty machine tools' remanufacturing,and the results are modified by the comprehensive adjustment method,which makes the evaluation results accord with the standard of human conventional thinking.Using the catastrophe progression method to establish the heavy duty machine tools' quantitative evaluation model,to evaluate the retired TK6916 type CNC floor milling-boring machine's remanufacturing.The evaluation process is simple,high quantification,the result is objective.

1.Introduction

Heavy duty machine tools are products that consume a great deal of resources, such as:DL250 heavy horizontal lathe,only its machine tool bed used dozens of tons of gray cast iron [1].The electrical components of heavy duty machine tools are updated speed, easy aging,but the iron castings which accounts for 60%~70% of the total weight of the machine tools is different[2,3].Though remanufacturing,a great deal of casting and processing costs can be saved.Machine tool manufacturing enterprises can make the business from manufacture,sale and service to the EOL stage of the product life cycle management.Remanufacturing of heavy scrap machine tools save lots of resources and energy,reduce the environmental burden caused by the production of new machine tools[4].

Remanufacturability is the premise to evaluate whether the product can be remanufactured.Ghazall and Murat proposed an optimal decision model for the remanufacturing scheme of waste components and parts[5].Some researchers from the residual life prediction of waste machine tool parts and the fatigue life[6].Remanufacturing evaluation is a systematic evaluation process,a scientific quantitative research work.Commonly used evaluation methods are analytic hierarchy process (AHP), grey relational grade method,etc.AHP is widely used because of its simplicity, systematicness, practicability.However,AHP needs to give weight to the evaluation index,the quantitative data is few and the qualitative component is more.In this paper,the method of catastrophe progression is used to evaluate the remanufacturing performance of heavy duty machine tools which absorbs the advantages of AHP and combines qualitative and quantitative analysis on the basis of considering relative importance of evaluation indexes. In order to overcome the shortcomings of the catastrophe progression method,the evaluation results are too high,the results are modified to accordance with people's regular judgment thinking.

2 The remanufacturing evaluation of modified catastrophe progression method

2.1 Modified catastrophe progression method

The catastrophe progression method takes the catastrophe theory as the theoretical basis, and the research object of the catastrophe model is the potential function.The main feature of the catastrophe progression method is that it decomposes the system's overall objectives at first,and then decomposes the fuzzy membership functions by combining the catastrophe theory with the fuzzy mathematics,the quantitative operation is performed by normalization formula,and finally normalized to a parameter,so as to evaluate[7].Common types of mutant systems are cusp catastrophe,swallowtail catastrophe,and so on.The normalization formulas of their mathematical models are:cusp catastrophe: $x_u=(u)^{\frac{1}{2}}$, $x_v=(v)^{\frac{1}{3}}$, $T=\min\left((x_u)^{\frac{1}{2}},(x_v)^{\frac{1}{3}}\right)$, swallowtail catastrophe: $x_u=(u)^{\frac{1}{2}}$, $x_v=(v)^{\frac{1}{3}}$, $x_w=(w)^{\frac{1}{4}}$, $T=\min\left((x_u)^{\frac{1}{2}},(x_v)^{\frac{1}{3}},(x_w)^{\frac{1}{4}}\right)$.

Catastrophe progression method are widely applied in many fields such as natural science,mechanical engineering and so on[8].In this paper,the method of comprehensive adjustment is used to modified the evaluation results.On the basis of understanding the evaluation system of catastrophe progression method, the underlying control variable X_i is set a certain value.Y_i is the superior evaluation index value.X_i,Y_i are as the rating scale.The scale is shown in Table 1.

Table 1 The scale of initial integrated value evaluation

X_i	0	0.20	0.40	0.60	0.65	0.70	0.75	0.80	0.85	0.90	0.95	1
Y_i	Y_0	Y_1	Y_2	Y_3	Y_4	Y_5	Y_6	Y_7	Y_8	Y_9	Y_{10}	Y_{11}

When the mutation series obtained comprehensive evaluation index $Y\in(Y_i, Y_{i+1})$,the evaluation of the modified X' should meet $X'\in(X_i, X_{i+1})$.X' can be obtained as in equation (1):

$$\frac{X'-X_k}{X_{k+1}-X_k}=\frac{Y-Y_k}{Y_{k+1}-Y_k} \tag{1}$$

2.2 The model of remanufacturing evaluation

The second class indicator of remanufacturing assessment model is divided into the environmental index (T_e), the economic index(T_i) and the technical index(T_s). T_e can be thought over from the material recycling rate, the type of material and so on.T_i should include the machine tool manufacturers purchase profit, economic benfits,environmental and social aspects.But because the difficulty in obtaining data and other factors, only consider the first two items.T_s is based on the process of remanufacturing.In fact,the index of disassemble closes to 1 and is discarded.Construct a remanufacturing evaluation model for heavy duty machine tools, as shown in Fig 1.The importance of the control variables is different in the mutation model, which is determined by itself.

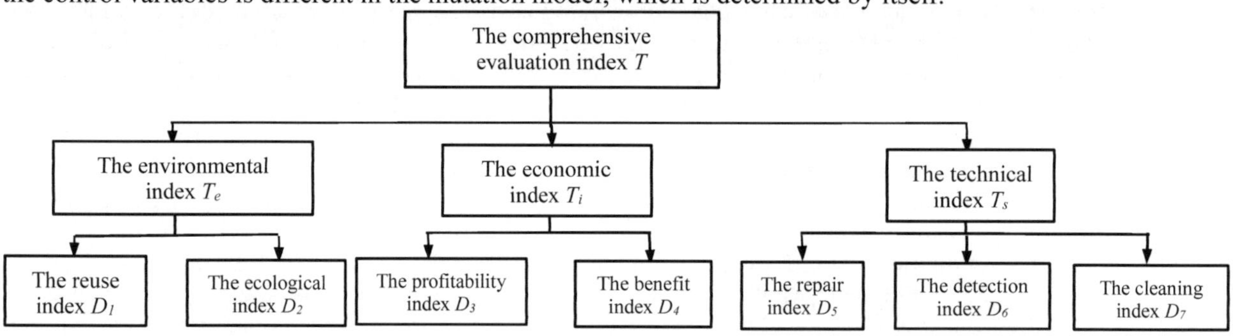

Fig 1. The remanufacturing evaluation model of heavy duty machine tools

2.3 The index of remanufacturing evaluation

2.3.1 The environmental index

Compared with the new manufacturing machine tools, the more materials can be saved, the more feasible the remanufacture is.And the quality and the type of materials will affect the environmental feasibility.So T_e can be obtained through the following process.The index (D_1)is mainly affected by the ratio of the total abandon parts quality (M_q) and the machine tool weight(M),as in equation (2).

$$D_1 = 1 - \frac{m_q}{M} \tag{2}$$

The ecological index reflects the environment effects of the materials. Components and parts with the different materials can influence the energy consumption, discharge, reuse,etc. The ecological index of heavy duty machine tools common using materials is shown in Table 2[9].Through the resuing weight of different materials (m_i) and its ecological index (EI_i) to get the ecological index of the whole machine (D_2).M_i is the weight of the whole machine parts using the material i.

Table 2. The main ecological index of heavy duty machine tools

Materials	Aluminium alloy	Copper alloy	Carbon steel	Cast iron	Plastic
Ecological index	10.0-18.0	60.0-85.0	4.0-4.3	3.0-10.0	4.2-4.3

$$D_2 = \min\left(\frac{\sum EI_{i\max} m_i}{\sum EI_{i\max} M_i}, \frac{\sum EI_{i\min} m_i}{\sum EI_{i\min} M_i}\right) \tag{3}$$

2.3.2 The economic index

This paper use the manufacturers profitability index (D_3) and users' benefit index (D_4) to evaluate the economic feasibility of heavy duty machine tools.

Remanufacturing cost mainly consists of three parts:recycling cost (C_1),remanufacturing cost (C_2),other cost (C_3).If the price of remanufactured products is the 40%-70% of the new products with the same performance,the manufacturers will profit[10].Therefore,we can determine the price of remanufacturing heavy duty machine tools (C).The manufacturer profitability index (D_3) is obtained as in equation (4).

$$D_3 = \frac{C - \sum_{j=1}^{3} C_j}{\sum_{j=1}^{3} C_j} \tag{4}$$

The users' benefits index (D_4) is influenced by the price of remanufacturing heavy duty machine tools (C),the price of new machine (C_0),the service life of remanufacturing heavy duty machine tools (N),and the service life of new machine (N_0).The users' benefits index (D_4) is obtained as in equation (5).

$$D_4 = \left(1 - \frac{C}{C_0}\right) \times \frac{N}{N_0} \tag{5}$$

2.3.3 The technical index

Common cleaning methods are blowing,wiping,baking,chemical cleaning, ultrasonic cleaning, etc.The degree of difficulty in the distribution of these cleaning methods is {0.2, 0.4, 0.5, 0.7, 0.9}[7].θ_j is the degree of the cleaning method j.N_j is the number of the cleaning method j is used.D_7 is the index of cleaning.As shown in equation (6).

$$D_7 = 1 - \frac{\sum_{j=1}^{5} N_j \theta_j}{N_j} \qquad (j=1,2,3,4,5) \tag{6}$$

The smaller the ratio of m_d and M is,the greater the index of test will be.m_d is weight of the non reusable class.M is the weight of the whole .The index of test (D_6) is obtained as in equation (7).

$$D_6 = 1 - \frac{m_d}{M} \tag{7}$$

Different types of components,because of using environment,load and other factors,will lead to a different rate of repair.D_5 is the index of repair,and can be calculated by formula(8).m_r is the weight of components and parts with no value of remanufacturing.M_r is the weight of being tested.

$$D_5 = 1 - \frac{m_r}{M_r} \tag{8}$$

2.4 Comprehensive evaluation

D_1 and D_2 constitute cusp catastrophe model,D_3 and D_4 constitute cusp catastrophe model, and D_5,D_6 and D_7 constitute swallowtail catastrophe model.T_e, T_i and T_s constitute swallowtail catastrophe model.Indicators at each level can not be complementary between each other,combining the dovetail catastrophe model, the normalized formula and select the minimum. T is the comprehensive evaluation index of heavy duty machine tool remanufacturing which needs to be modified by (1) to obtain the regular evaluation results T'.At present, the regular scoring standard is 1, and the pass is 0.6.According to the value of T',make the following judgement:when $T'\geq 0.6$,retired heavy duty machine tool is can be remanufactured.And the greater the T' is,the more the heavy duty machine tool can be remanufactured.

3 Example analysis

This paper studies an TK6916 type CNC floor milling-boring machine with a service life of more than 10 years in Wuhan heavy duty machine tool group corporation.Obtaining the relevant data from the manufacturers and according to assessment process,we get the following values in Table 3.

Table 3. Remanufacturing evaluation index

Secondary indicators	Environmental feasibility T_e		Economic feasibility T_i		Technical feasibility T_s		
Thirdary indicators	Reuse index D_1	Ecological index D_2	The profitability index D_3	The benefit index D_4	Cleaning index D_5	Test index D_6	Repair index D_7
Value	0.841	0.910	0.692	0.585	0.565	0.889	0.876

According to the normalization formula of the catastrophe progression method, T_e is 0.917, T_i is 0.832, T_s is 0.867, T is 0.941.According to the correction principle and formula (1), the T is modified, and T' is 0.773.That is to say, the CNC floor milling-boring machine has higher manufacturability.

4 Summary

This paper uses catastrophe progression method to comprehensive quantitative evaluate the remanufacturing of heavy duty machine tools.Reduce the qualitative components caused by artificial weighting,the quantitative component of the synthetic analysis was increased.Considering that the catastrophe progression method can lead to larger evaluation result in the normalization process,comprehensive evaluation method is used to modify the evaluation results,to conform the habits of people's assessment of the pros and cons.Based on catastrophe theory, the remanufacturing evaluation model of heavy duty machine tools is established,and calculates the various levels of the remanufacturing evaluation model according to the characteristics of heavy duty machine tools.The remanufacturing performance of the retired TK6916 type CNC floor milling-boring machine is evaluated by using the modified catastrophe progression method proposed in this paper.The evaluation of the remanufacturing is affected by various uncertainties,but the evaluation results of the method proposed in this paper can provide a decision basis for the remanufacturing process of heavy duty machine tools, and the evaluation results are within the acceptable range.

Acknowledgments

The paper is supported by the National Sci-Tech Support Plan of China under Grant Nos.2015BAF32B03.

References

[1] Zhang Mingqing.Research on the manufacturing process of long lathe bed of heavy duty machine tools[D].Hunan University,2014.
[2] Du Yanbin.Study on Evaluation and Redesign Methods of Machine Tool Remanufacturing[D].Chongqing University,2012.

IOP Conf. Series: Materials Science and Engineering **265** (2017) 012029 doi:10.1088/1757-899X/265/1/012029

[3] Zhao Jiaxue.Dynamic Stability and Assembly Process Method of Heavy Duty Machine Tool[D].Harbin University of Science and Technology,2015.

[4] Yawei Hu,Shujie Liu and Hongchao Zhang.Remanufacturing decision based on RUL assessment[J].*Procedia CIRP*,2015,**29**:764-768.

[5] Ghazalli Z and Murata A.Development of an AHP-CBR evaluation system for remanufacturing: End-of-life selection strategy [J]. *International Journal of Sustainable Engineering*,2011, **4(1)**:2-15.

[6] Liu Guangyu.Research ON Residual Life Prediction Technology For Remanufactring Of Machine Tool Parts[D].Harbin Institute of Technology,2014.

[7] Li Ming Xu.Risk Assessment and Study on Prevention of Debris Flows near the Dam of Wudongde Hydropower Station Based on Catastrophe Theory[D].Jilin University,2013.

[8] Jun Wang,Guanhua Huang,Jiusheng Li,Jianhua Zheng,Quanzhong Huang,Haijun Liu. Effect of soil moisture-based furrow irrigation scheduling on melon (Cucumis melo L.) yield and quality in an arid region of Northwest China[J]. *Agricultural Water Management*,2016,**179**:167-176.

[9] Pengjia W , Yongxian L , S.K. Ong and A.Y.C. Nee .Modular Design of Machine Tools to Facilitate Design for Disassembly and Remanufacturing[J].*Procedia CIRP*,2014,**(15)** :443-448.

[10] Du Yanbin, Cao Huajun, Liu Fei, Yan Ping and Li Congbo.Evaluation of machine tool remanufacturing scheme based on entropy weight and AHP[J].*Computer Integrated Manufacturing Systems*,2011,**(01)**:84-88.

Effect of silicon source and carbon source on the morphology of SiC powders

Zhao WANG[1,*], Shu-lin WANG[1,2], Jia-you JI[1,2] and Man XU[1,*]

[1]School of materials science and Engineering, Wuhan Institute of Technology, Wuhan, Hubei, 430073;
[2]Engineering Research Center of Environmental Materials and Membrane Technology of Hubei province, Wuhan, Hubei, 430073;

*Corresponding author e-mail address: 1733784513@qq.com

Abstract. Silicon carbide (SiC) powders were prepared by carbothermal reduction method with silica fume and carbon-white as the silicon source; glucose, phenolic resin, polyvinyl pyrrolidone (PVP) as carbon source. The effects of reaction temperature, silicon source and carbon source on the morphology of silicon carbide powder were investigated. The result shows that: The optimum temperature for preparation of silicon carbide powder was 1400 °C; When the glucose was used as carbon source, the morphology of SiC powder was greatly affected by the kinds of silicon sources; When the phenolic resin and PVP were used as carbon source, the morphology of SiC powder was greatly influenced by the carbon source.

1. Introduction

Silicon carbide (SiC) ceramic has excellent mechanical properties, oxidation resistance, wear resistance, thermal stability, thermal shock resistance and chemical corrosion resistance, and the thermal expansion coefficient is small, thermal conductivity is large[1-6]. Different size and morphology of SiC powders can be used to prepare different types of SiC ceramic products. Therefore, the study of the influence factors of SiC powder morphology has high practical significance and market value.

In this paper, SiC powder was prepared by carbothermal reduction method with different silicon source and different carbon source, the effects of reaction temperature, silicon source and carbon source on the morphology of silicon carbide powder were investigated. It has a certain guiding significance to the actual production of silicon carbide powder.

2. Experimental

2.1. Reagents and Instruments

Reagents: silica fume, Tech, Gansu Sanxin silicon industry limited company; carbon-white, Tech, Weifang Sanjia Chemical Co. Ltd. ; glucose, AR, Chongqing Heping Pharmaceutical Co. Ltd. ; phenolic resin, AR, Tianjin United Chemical Industry Co. Ltd. ; polyvinyl pyrrolidone (PVP), AR, Henan Zheng Yao chemical products Co. Ltd.

Instruments: Analytical balance (08F106-42), Electrothermal constant temperature blower drying box(DHG-9076A), Tubefurnace(SK_2-6-12A), Muffle furnace(LHT807GN4G), Horizontal vacuum sintering furnace(ZT-45-22W).

2.2. Sample Preparation

Frist of all, the carbon source was dissolved in distilled water, and then added a certain amount of silicon source, ball milled the slurry at high speed for 2 hours. Then put the slurry in the oven under 100 °C for 24 h. After the slurry was dried and pulverized, put the powder into a tubular furnace heated to 650 °C for 2 h under nitrogen atmosphere to prepare precursor, the heating rate was 5 °C /min. The precursor was heated to reaction tempeture under vacuum for 2 h. The synthesized powders were washed with hydrofluoric acid (HF) to remove unreacted silica and other impurities, then heated the powders to 650 °C in atmosphere for 2 h to remove the residual carbon in the product. Finally, the SiC powder could be obtained by washing, filtering and drying.

2.3. Sample characterization

The phase composition and crystallinity of the materials were characterized by powder XD-5A x-ray diffraction（XRD）which was produced by Japan Island Company.
Microstructures of the products were observed on a JSM-5510LV Scanning Electron Microscope（SEM）which was produced by Japan.

3. Results and Discussion

3.1. Effect of reaction temperature on the morphology of silicon carbide

The X- ray diffraction (XRD) analysis of the product was carried out at as shown in Figure 1. The reaction conditions was that: The micro silica fume used as the silicon source and glucose used as the carbon source under the 1300 °C, 1400 °C, and 1500 °C for 2 h.

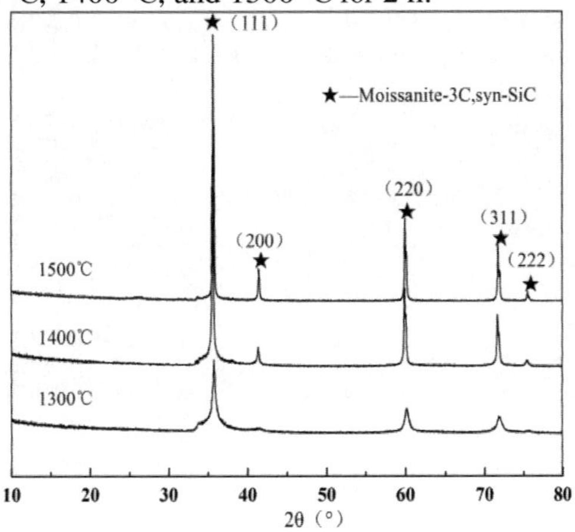

Fig.1 XRD patterns of the products at different temperatures

At 1300 °C, 1400 °C, and 1500 °C calcined 2 h, the obtained products were 3C SiC, the crystalline SiC belongs to β-SiC. From the intensity of diffraction peaks of 3 kinds of samples, the diffraction peak of the product at 1500 °C was the strongest, followed by 1400 °C, and the diffraction peak of the 1300 °C product was the weakest. At 1300 °C, the crystallinity of the product was not high, and with the increase of temperature, the diffraction peak became sharp. It was known that the crystallinity of the product was becoming better, and the crystal was gradually developed.
The SEM test on three kinds of temperature on the morphology of the products were shown in Figure 2.

Fig.2 Scanning electron microscope（SEM）photographs of the products at different temperatures (a)the product at 1300 °C；(b)the product at 1400 °C；(c)the product at 1500 °C

As shown in Fig.2(a), when the reaction temperature was 1300 °C, the sample consisted of some amorphous SiC particles , mutual adhesion between particles, particle size wss about 100 nm. It can be seen from Fig. 2(b) that the sample particles were obvious, particles were connected to each other, high sphericity, single particle size was about 200 nm. It can be seen from Fig. 2(c) that the silicon carbide particles obviously grew up , the large particles showed sharp edges and corners, the morphology of some small particles were still nearly spherical. In contrast to 1300 °C and 1400 °C, the particles grew obviously and the sphericity became worse. At this temperature, the SiC crystal was the most complete and the particles were growing up, the above results were in agreement with the XRD data. Therefore, in this experiment, the optimum temperature for preparation of silicon carbide powder was 1400 °C.

3.2. Effect of silicon sources on silicon carbide morphology

The experiment used silica fume and carbon-white two silicon source, used glucose as carbon source, the SiC powder was prepared under the temperature 1400 °C for 2 h , and conducted SEM tests on raw materials and products, the test results were shown in Figure 3.

silica fume	SiC	carbon-white	SiC

Fig.3 Scanning electron microscope（SEM）photographs of the silicon sources and products

As shown in Fig.3, when the glucose was used as carbon source, the morphology of SiC powder was close to that of silicon source, this may be because during the experiment, the glucose could be completely soluble in water, the silicon source and the carbon source could be mixed evenly. In the carbonization process, the carbon would be wrapped in silicon source around to form a coating layer, under the reaction temperature, the silicon source gradually turned into a molten silicon source, and the carbon was gradually dissolved into the silicon source to form silicon carbide powder. In the process, silicon source to melt and the carbon source was gradually dissolved, leading to the the reaction to continue, eventually lead to silicon carbide inherited silicon source morphology.

3.3. Effect of different carbon sources on the morphology of silicon carbide

The experiment used silica fume as silicon source, changed the types of carbon source to study the effects of different carbon sources on the morphology of silicon carbide. Conducted SEM test on the product and the test results were shown in Figure 4.

glucose as carbon source	phenolic resin as carbon source	PVP as carbon source

Fig.4 Scanning electron microscope（SEM）photographs of products with different carbon source

The silicon carbide powder of different carbon sources were fabricated by SEM analysis, it can be seen that: Different carbon sources lead to different morphologies of SiC, when glucose was used as carbon source, SiC was spherical and granular; when phenolic resin was used as carbon source and SiC was flocculent; when PVP was used as carbon source, SiC was amorphous powder. This may be due to phenolic resin and PVP as carbon source, its solubility in water ws not as good as glucose, resulting in mixed carbon source and silicon source was not uniform, in the process of carbonization, the carbon source could not completely cover the silicon source. In the reaction process, the silicon source was melted and flowed in the reaction system, a portion of the silicon source flowed to the surface of carbon source and reacted with the carbon to form SiC. At present, the silica fume unable to maintain its original shape, so the product could not inherit the morphology of the silicon source, but the shape of the carbon source determined the morphology of the product.

4. Conclusions

(1) The optimum temperature for preparation of silicon carbide powder was 1400 ^{o}C;
(2) When the glucose was used as carbon source, the morphology of silicon carbide inherits the morphology of silicon source;
(3) When the phenolic resin and PVP were used as carbon source, the morphology of silicon carbide is flocculent, and the morphology of silicon source is not well inherited.

Acknowledgments

This work was supported by the Postgraduate Innovation Education Fund (CX2016013), Hubei Province Science and Technology Support Project (Grant No.2014BAA102) and Project of Technology Innovationin Hubei Province (Grant No.2016ACA160).

References

[1] A. Najafi, F.Golestani-Fard, H.R.Rezaie, N.Ehsani, J.Sol-Gel Sci.Technol. 59, 205-214(2011).
[2] Solomah, A. C., Reichert, W., Rondinella, V., Esposito, L., & Toscano, E. (2010). Journal of the American Ceramic Society,73(3), 740-743.
[3] Perham, T. Office of Scientific & Technical Information Technical Reports (1996).
[4] Chuan-Jun, T. U., et al. Transactions of Nonferrous Metals Society of China 25.3(2015):856-862.
[5] Okuno, Akiyasu, and M. Watanabe. US, US 4946807 A. 1990.
[6] Lee, Jin Seok, D. M. Choi, and S. C. Choi. Solid State Phenomena 124-126(2007):747-750.

IOP Conf. Series: Materials Science and Engineering **265** (2017) 012031 doi:10.1088/1757-899X/265/1/012031

Study on extrusion process of SiC ceramic matrix

Xiao-Yuan DAI[1], Fan SHEN[1], Jia-You JI[1,2],Shu-Ling WANG[1,2,] Man XU[1,2,*]

[1]School of materials science and Engineering, Wuhan Institute of Technology, Wuhan, Hubei, 430073;

[2]Engineering Research Center of Environmental Materials and Membrane Technology of Hubei province, Wuhan, Hubei, 430073

*Corresponding author e-mail address:1733784513@qq.com

Abstract. In this thesis, the extrusion process of SiC ceramic matrix has been systematically studied.The effect of different cellulose content on the flexural strength and pore size distribution of SiC matrix was discussed.Reselts show that with the increase of cellulose content, the flexural strength decreased.The pore size distribution in the sample was 1um-4um, and the 1um-2um concentration was more concentrated. It is found that the cellulose content has little effect on the pore size distribution.When the cellulose content is 7%, the flexural strength of the sample is 40.9Mpa. At this time, the mechanical properties of the sample are the strongest.

1. Introduction

The porous silicon carbide ceramics have a large surface area, hydrophilic, excellent performance, high temperature, resistance to acid corrosion, light, etc., it is widely used in various fields catalyst industry, metallurgical industry, sewage treatment, car exhaust emissions treatment[1].Now the market is mainly recrystallization[2-4] method of preparing porous silicon carbide ceramic, but its complex manufacturing process, high production cost, is not conducive to a wide range of application porous ceramic silicon carbide.Mullite and silicon carbide has good chemical compatibility, similar chemical creep resistance, can be better at a low temperature silicon carbide are bonded together using mullite bonded silicon carbide, silicon carbide porous ceramic preparation can be achieved low-temperature sintering. Therefore, this paper uses mullite bonded silicon carbide [5]to prepare silicon carbide porous ceramic.not only greatly reduce the cost but also have relatively simple preparation process.

2.Experimental part

2.1.Experimental material

Table 1. Experimental raw materials and experimental equipment

Experimental material name	Experimental equipment
silicon carbide	ball mill tank
silica fume	agitator
alumina	tube type resistance furnace
polyethylene glycol	ceramic tube extrusion machine
glycerol	muffle furnace
oleic acid	electric blower drying box
hydroxypropyl methylcellulose	

2.2.Experimental method

(1) Mix Pour the weighing ingredients into the blender in turn, and stir in the mixture for a certain amount of time. Finally, add the aqueous solution of polyethylene glycol to a small amount of water, stirring for 20 minutes;

(2) Clay preparation The mixture was stirred in the ceramic tube extruding machine pugging three times, make all kinds of raw materials full contact, exclude mud inside air, avoid extrusion products within the air bubbles appear on the surface;

(3) Stale The stirred mixture contains a variety of ingredients that are confused and contain a lot of air, which causes pores on the surface of the product when it is extruded. Therefore, the mixture would be a good mix of practice mud wrap, stale 12h, so that the sample of hydroxypropyl methyl cellulose can better react with water, enhance the plasticity of paste viscosity is reduced, extrusion molding for samples;

(4) Squeeze Select the desired shape of the product mold, and then use the ceramic pipe extrusion machine extrusion products;

(5) Dry The extruded products first placed in the air naturally dry for 12h, and then placed in the oven at 100 ℃ temperature drying, the time is 2h;

(6) Burn Cut the dried good products into slices, smooth them, and place them in the resistance furnace and pre burning them to remove the organic plasticizer in the sample.

2.3.Experimental characterization

2.3.1.Flexural strength test. The bending strength can be called bending strength, that is, the maximum pressure occurs when the product is broken. The flexural strength of SiC ceramics is closely related to the distribution of bubbles inside the product, the complete mixing of the particles, porosity and pore size distribution. In this experiment, the bending strength of the products was measured by three point method. The products are firing ultrasonic cleaning to remove the product surface of carbon black, and then dried in an oven at a temperature of 100℃ for 2 hours. Finally, the flexural strength of the product is measured by a computer controlled insulation material testing machine. During the experiment, the length, width and thickness of the product were measured with a vernier caliper, and then the vertical pressure was applied to the middle part of the product until the product was broken, and the data were recorded.

2.3.2.Detection of pore size distribution. In this experiment, the pore size distribution of the test article was measured by the bubble point method, the test sample was placed in a solution with a large wetting angle so that the pores in the sample to be measured were filled with the infiltration solution.The diameter of the through-hole of the sample to be measured is calculated by measuring the size and flow rate of the gas flow rate of the sample to be measured (the test results characterize the flux of the through-hole, which is more meaningful than the mercury-mercury meter data), using the equipment for the POROMETER pore size distribution tester (Germany, Porolux Nv company).

3.Results and discussion

3.1.Influence of cellulose content on folding strength

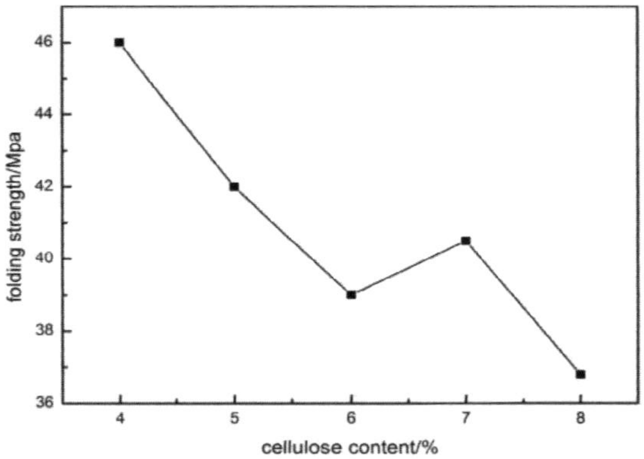

Figure 1. The strength of the anti - folding strength varies with the content of cellulose

As can be seen from Fig 1,with the increase of cellulose, the flexural strength of products will gradually decrease.When cellulose is 4%, the flexural strength of the product is highest, reach to 46Mpa. When the cellulose was added to 5%, the flexural strength gradually decreased to 42.1Mpa, and the amount of cellulose added to 6% continued, then the flexural strength continued to decrease and reached a minimum peak value of 38.9Mpa.The 7% products in the fiber, the flexural strength increases suddenly,due to the hydration reaction of cellulose and water more completely, the products of cellulose is not too much, leaving the pore is relatively small, so the flexural strength will suddenly increase. When cellulose was added to 8%, the flexural strength decreased and reached a minimum 36.8Mpa, when the product was easy to break.In the process of mixing, the cellulose added will volatilize in the pre burning process, and the pore left after volatilization will reduce the flexural strength of the products and break easily. Therefore, the flexural strength will decrease with the increase of cellulose.

3.2.Effect of cellulose content on pore size distribution

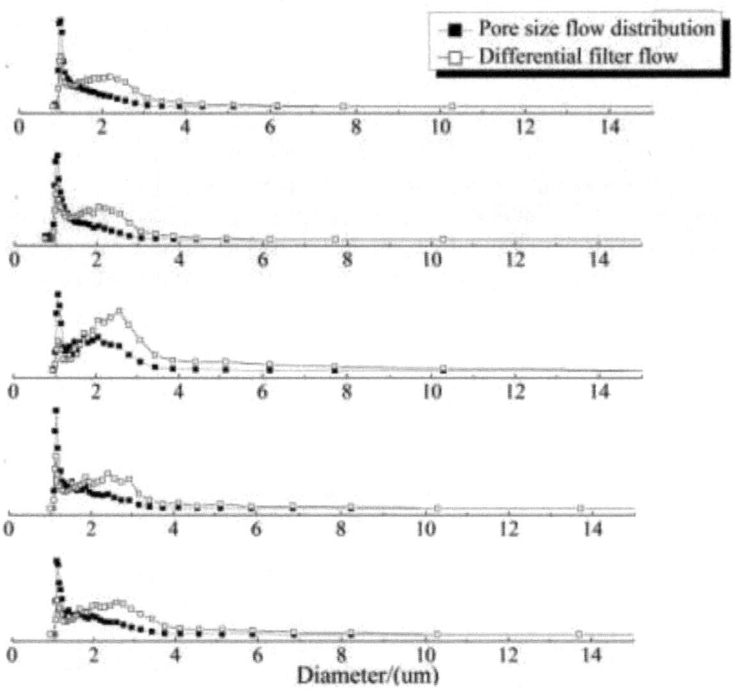

Figure 2. The strength of the pore size distribution with the content of cellulose

Fig 2 is the pore size distribution and volume distribution of samples with different fiber content, it can be seen from the figure, with the increase of cellulose content was smaller pore size distribution and gas flow distribution in the sample, the sample aperture roughly distributed between 1um-4um,and 1um-2um distribution is relatively concentrated, When the airflow passes through the sample, a less part of the air flow passes through 1um-2μm holes between , and the other air streams pass through larger holes between 2um-4um, although the pore size is mainly concentrated between 1um-2um, but the airflow is mainly from 2um-4um The hole is passed so that the air flow of the sample is mainly due to the large pore distribution of 1um-2um pores and the distribution of less than 2um-4um between the large pore contribution.

4.Conclusion

The addition of cellulose has a great impact on the performance of the product. With the increase of the amount of the cellulose, the porosity of the product gradually increases, the distribution of the pore size is gradually concentrated, the flexural strength decreases first and then increases trend. When the cellulose content is 7%, the flexural strength of the sample reaches 40.9Mpa, and the mechanical properties of the sample are the best.

Acknowledgement
This work was supported by Hubei Province Science and Technology Support Project(Grant No.2014BAA102) and Project of Technology Innovation in Hubei Province(Grant No.2016ACA160)

References

[1] Cao Z, Wei B. *Energy Environ Sci.* **6** 3183.
[2] Serre C, Calvo Barrio L, PerezRodriguez A, et al. *J Appl Phys.* **79** 6907.
[3] Bohn, H G, Williams, J M, McHargue, C J,et al. *J Mater Res.* **2** 0716.
[4] Serre C, Calvo B L, Perez R A, et al. *J Appl Phys.* **79** 0713.
[5] Ji X L, Xu F, Chen H Y. *Adv Mat Res.* **79** 1983.

IOP Conf. Series: Materials Science and Engineering **265** (2017) 012032 doi:10.1088/1757-899X/265/1/012032

Synthesis of Highly Ordered Macroporous Titania Using TiCl4 Hydrolysis

Chong Wang, Cancan Yang, Mei Yu*, Zhe Yu.

Biomedical Microdevices Research Laboratory, Shenzhen Institutes of Advanced Technology, Chinese Academy of Sciences, 1068 Xueyuan Avenue, Shenzhen 518055, China

* Corresponding Author E-mail: mei.yu@siat.ac.cn

Abstract: The high quality ordered macroporous titania has been successfully synthesized with the hydrolysis of titanium tetrachloride (TiCl4) by heating the hydrochloric acid solution of TiCl4 which filled in polystyrene templates. In this method, there is no need of filling the interspaces of polystyrene template repeatedly and no introduction of organic impurities either. The best conditions of synthesizing ordered macroporous titania are found that the pH value is 2.0～2.5, the filling time is 3h and the calcination temperature for the filled templates is 500℃. The macroporous titania synthesized show highly ordered structure and obvious photonic bandgap.

1. Introduction

The ordered macroporous titania has attracted considerable interest due to their large specific surface area, high porosity and high order degree. It is widely used as absorption materials [1], battery materials, macromolecular catalysts and especially a kind of high-quality photonic crystals[2, 3].Because of the periodical distribution of the dielectric constant, the ordered macroporous titania is extensively used in the fields of optical wave-guide[4], low-threshold lasers, photonic chipolystyrene, optical communications and so on.

At present, the template method[5-7], regarded as an economical and well size-controlled way, is commonly used to synthesize the ordered macroporous titania. However, to fill the template interspaces with the titania effectively is always arduous. Mostly, the Sol-gel [8-12] method is used to fill the template interspaces. In the Sol-gel method, the collosol is prepared with butyl titanate first, then the template is immersed in the collosol to fill the interspaces, at last the template substance is removed by pyrolysis. The collosol has formidable access to the template interspaces, thus the filling of the collosol must be repeated several times, which affects the structural integrity of the ordered macroporous titania. Moreover, a mass of organic impurities is introduced in the process, which might affect the structural uniform of the macroporous titania. In addition, there are some other methods of filling the template interspaces, such as chemical vapor deposition and electrochemical deposition method [13]: the former with even filling and good controlled filling-rate but high experimental conditions and complex operation while the latter inapplicable to the titania materials.

For the purpose of filling the template interspaces more effectively, we have adopted a novel approach. First we prepared the polystyrene template, and then put the template into the hydrochloric acid solution of titanium tetrachloride. Adjusting the pH value of titanium tetrachloride solution by heating

IOP Conf. Series: Materials Science and Engineering **265** (2017) 012032 doi:10.1088/1757-899X/265/1/012032

to make hydrochloride volatilize, the titanium tetrachloride in the template interspaces began to hydrolyze slowly at a certain pH value, until to fill the template interspaces uniformly by the hydrated titania. At last, the polystyrene template was removed by pyrolysis and the hydrated titania turned to titania at one time. This method is a simple and effective way to synthesize the ordered macroporous titania. There is no organic impurities introduced or no need of repeated filling process. With the uniform structure and high order rate, the macroporous titania synthesized has an obvious photonic bandgap in the infrared area.

2. Experimental

Firstly, the polystyrene microspheres were synthesized by non-soap emulsion polymerization method, and then the polystyrene templates were assembled by the vertical deposition method [14]. The vertical deposition was carried on with a temperature at 65℃ and a humidity at 90%. At last, the polystyrene templates were heated for 5 hours at 85℃.

The synthesis steps of the ordered macroporous titania are as follows: 1) the hydrochloric acid solution of titanium tetrachloride was prepared: the concentration of titanium tetrachloride was 0.8mol/L and the concentration of hydrochloric acid was 2mol/L; 2) the prepared polystyrene templates were immersed in the hydrochloric acid solution of titanium tetrachloride and heated at 80 ℃, in the entire process to maintain the hydrochloric acid solution of titanium tetrachloride in the constant temperature and simultaneously using the hydrochloric acid solution (1mol/L) to adjust the pH value between 2.0 ~ 2.5, until to fill the template interspaces uniformly by the hydrated titania after 3 hours; 3) the polystyrene templates were removed out and drying at room temperature; 4) the filled templates were calcined at 500 ℃ for 30 minutes in high-temperature furnace. Finally, the ordered macroporous titania was prepared.

Figure 1. shows the process of the ordered macroporous titania. The interspaces of polystyrene templates could be filled with the hydrochloric acid solution of titanium tetrachloride at first. Then by heating, hydrogen chloride in the solution was gradually volatilized, making the pH value of the solution rising. As a higher pH value, titanium tetrachloride began the process of hydrolysis and generated hydrated titania. Finally, after calcining at 500℃, polystyrene was completely decomposed meanwhile titania was generated by dehydration of hydrated titania.

SEM images were taken withscanning electron microscope S440 produced by Leica Cambridge LTD. (voltage=25Kv and current=200pA). Raman spectra were characterized with laser Raman spectrometer InVia produced by Renishaw corporation. Optical absorption spectrum were recorded on a UV-Vis-NIR spectrometerUV-3150 produced by Shimadzu corporation.

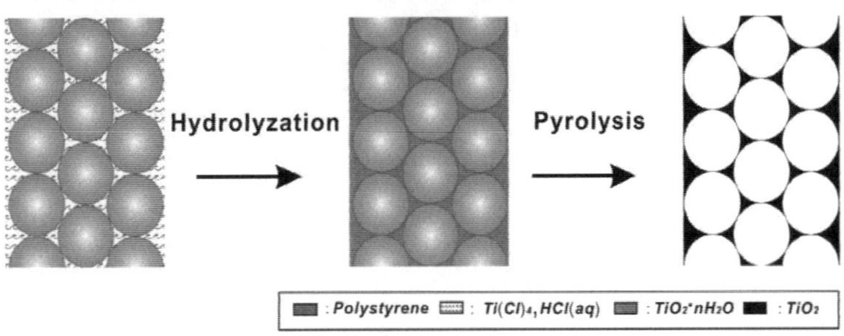

Figure 1. The process schematic diagram of synthesis of ordered macroporous titania

3. Results and discussion

IOP Conf. Series: Materials Science and Engineering **265** (2017) 012032 doi:10.1088/1757-899X/265/1/012032

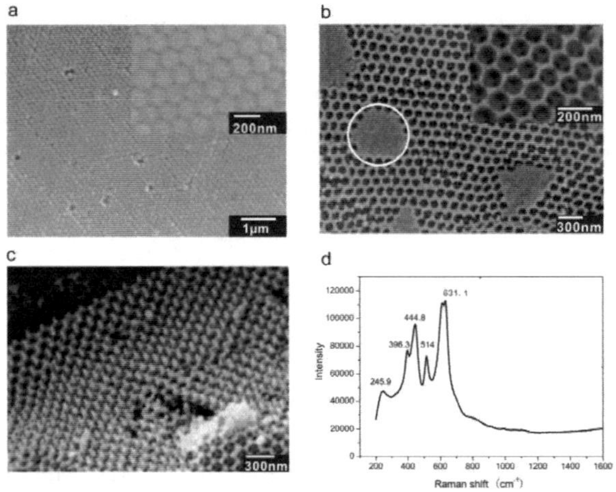

Figure 2. SEM image of (a) PS template, (b) the surface of ordered macroporous titania, (c) the section of ordered macroporous titania; (d) Raman spectrum of ordered macroporous titania.

The structure of ordered macroporous titania is determined by the polystyrene templates, so the quality of the template is indispensable. Figure 2(a) shows that the polystyrene microspheres in the templates, which has the size of 155nm±, were packed close and arranged in order.

Figure 2(b) shows that the average pore size of the macroporous titania is 130nm±, relatively contractive compared with the templates (155nm±), which resulted from the dehydration of hydrated titania in the process of calcination. Due to the defects in polystyrene template, there are also a few defects in the ordered macroporous titania. It is also show that the pores in the sample are distributed symmetrically in a highly ordered structure. In the circular marked area there are no pores, it is because the process of filling inside the templates has been finished already, nevertheless, the hydrated titania generated by titanium tetrachloride is still hydrolyzing and depositing on the surface and thus make the pores on the surface disappear. On the other hand, the ordered macroporous structure is still demonstrated inside the templates, as shown in Figure 2(c). SEM images imply that the structure of the macroporous titania is symmetrical and orderly from the surface to the bottom. Therefore, by the use of TiCl4 hydrolysis, the interfaces of templates can be filled with titania adequately and efficiently.Figure 2 (d) shows that the sample is the coexistence of anatase and rutile. In the vibration mode of anatase phase, the peak at 396.3cm-1 is B1g vibration mode; the peak at 514cm-1 is B2g vibration mode and the peak at 631.1cm-1 is Eg vibration mode. In the vibration mode of rutile phase, the peak at 245.9cm-1 is vibration mode of multi-phonon processes; the peak at 444.8cm-1 is Eg vibration mode and the peak at 613.5cm-1 is A1g vibration mode.

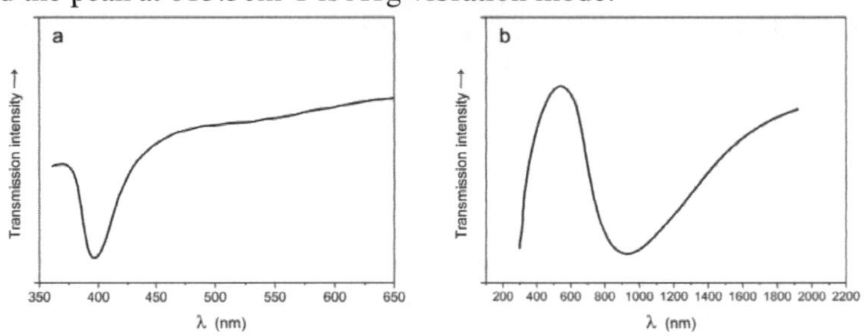

Figure 3. Optical absorption spectrum: (a) Polystyrene template and (b) Orderedmacroporous titania.

The pore structure of the macroporous titania prepared is highly ordered and periodically arranged in space, thus the titania and pore with different refractive indexes are also arranged alternately and periodically in space. Figure 3(b) shows that the bandwidth of the photonic bandgap of the macroporous titania sample is 200nm and the central wavelength is 925nm and the bandgap bandwidth of the template of polystyrene microspheres is 26nm and the central wavelength is 395nm in Figure 3(a). It reveals the bandgap position of the macroporous titania sample showing red shift and the bandgap width is widened obviously. Therefore, the highly ordered macroporous titania can be widely applied to the photonic bandgap materials.

4. Conclusion

In order to fill the polystyrene template interspaces more effectively with titania, a novel method is adopted for filling the interspaces of templates by the titanium tetrachloride hydrolysis. In this method, there is no need of filling the interspaces repeatedly and no introduction of organic impurities either. The best conditions for the synthesis are found: the pH value is 2.0～2.5; the filling time is 3h and the temperature of calcinations is 500℃. The Raman spectra show that the samples are the coexistence of anatase and rutile and the SEM images show that the samples are filled uniformly and the structure is orderly. Moreover, the optical absorption spectra illustrate that the samples, highly ordered, have an obvious photonic bandgap.

Acknowledgments

This work was supported by National Natural Science Foundation of China (61374013), National Program on Key Basic Research Project (973 Program, 2015CB755500), Shenzhen "Layout of Disciplines" Program on Basic Research Project (JCYJ20160331174854880), and China Postdoctoral Science Foundation 59th (162438).

References

[1] Schroden RC, Al-Daous M, Sokolov S. *Hybrid macroporous materials for heavy metal ion adsorption*. Mater Chem 2002;12: 3261–7.

[2] Yablonovitch E. *Inhibited spontaneous emission in solid-state physics and electronics.* Phys Rev Lett 1987, 58 : 2059–62.

[3] John S. *Strong localization of photons in certain disordered dielectric superlattices*. Phys Rev Lett 1987;58: 2486–9.

[4] Mekis A, Chen JC, Kurland I. *High transmission through sharp bends in photonic crystal waveguides*. Phys Rev Lett 1996; 77: 3787–90

[5] Zhou Z, Zhao XS. *Opal and inverse opal fabricated with a flow-controlled vertical deposition method.*Langmuir 2005; 21: 4717–23.

[6] Jiu J, Wang F, Sakamoto M, Takao J, Adachi M.*Performance of dye-sensitized solar cell based on nanocrystals TiO2 film prepared with mixed template method*. Sol Energy Mater Sol Cells 2005;87:77–86.

[7] Al-Daous MA, Stein A. *Preparation and Catalytic Evaluation of Macroporous Crystalline Sulfated Zirconium Dioxide Templated with Colloidal Crystals*. Chemistry of Materials 2003; 15: 2638–45.

[8] Yan HW, Smyrl WH, *Blandford CF, Holland BT. General synthesis of periodic macroporous solids by templated salt precipitation and chemical conversion*. ChemMater2000; 12: 1134–8.

[9] Abdelsalam ME, Bartlett PN, Baumberg JJ, *Preparation of arrays of isolated spherical cavities by self assembly of polystyrene spheres on self assembled prepatterned macroporous films*. Coyle S Adv Mater 2004; 16:90–3.

[10] Zukalová M, Zukal A, Kavan L, Nazeeruddin MK, Liska P, Gratzel M. *Organized mesoporous TiO2 films exhibiting greatly enhanced performance in dye-sensitized solar cells.* Nano Lett 2005; 5: 1789–92.

[11] Zhong Z, Yin Y, Gates B, Xia Y. *Opal and inverse opal fabricated with a flow-controlled vertical deposition method.*Adv Mater 2000; 12: 206–9.

[12] Wijnhoven JE, Vos WL. *Preparation of photonic crystals made of air spheres in titania.* Science 1998; 281: 802–4.

[13] Bartleet PN, Duinford T, *Ghanem MA. Templated electrochemical deposition of nanostructured macroporous PbO2*. Mater Chem 2002; 12: 3130–5.

[14] Dimitro AS, Nagayama K, *Langmuir. Continuous convective assembling of fine particles into two -dimensional arrays on solid surfaces* 1996; 12:1303–11.

IOP Conf. Series: Materials Science and Engineering **265** (2017) 012033 doi:10.1088/1757-899X/265/1/012033

Some methods of increasing the density of metal in order to increase him corrosion resistance

I V Chumanov[1], A N Anikeev[1], D V Sergeev[1] and A N Maltseva[2]

[1] Department of engineering and technology of materials, South Ural State University (national research university), 16 Turgenev Street, Zlatoust, Russia
[2] The Russian Research Institute of the Tube & Pipe Industries, Chelyabinsk, Russia

Abstract. Methods to increase the density of metal in order to increase its corrosion resistance in an aggressive environment are examined in the article. Two steel grades, differing in the content of alloying elements, increasing the resistance to corrosion are selected for the manufacture of experimental metallic materials. Two technologies are chosen as methods for increasing the density, and as a result, corrosion resistance, of the experimental materials obtained: the first is electroslag remelting with rotation of the consumable electrode, the second is centrifugal casting with modification. The microstructure of the metal becomes more homogeneous, the degree of metal refining from non-metallic inclusions increases, the rate of crystallization during metal smelting by the ESR method increases with rotation of the consumable electrode. When ingots are produced by the method of centrifugal casting, they are modified with dispersed WC and TiC particles, which increases the crystallization rate, increases the metal density, corrosion and mechanical properties. The evaluation of their corrosion resistance with the help of the autoclaved test complex "Cortest" is made after obtaining ingots by various technologies.

1. Introduction

The problem of the protection of metal structures from corrosion arose with all severity with the discovery and development of oil and gas fields in the composition of the produced fluid which contains hydrogen sulfide, carbon dioxide or their joint presence. About 350.000 km of commercial pipelines are located in Russia in operation, where 20.000 cases of leakage are observed every day, leading to oil losses and environmental pollution. In this case, about 90% of accidents occur due to corrosion damage to pipe material and stop valves. Thus, the problem of providing materials with resistance to corrosion in an aggressive environment is quite relevant today [1]. One of the important criteria is the density of the resulting metal to increase the resistance of metals to corrosion. Of the known technologies of metallurgical production, electroslag remelting (ESR) and centrifugal casting make it possible to produce a metal with an increased metal density index in comparison with other technologies [2-4]. In addition, a number of properties, including the density of the metal, can be increased by carrying out the ESR, by using the technology of rotating the consumable electrode [5]. The main layer acting on the metal during casting with centrifugal casting is the centrifugal force, which exceeds tens of times the gravity of the metal being poured. In the casting, a dense, fine-grained metal structure without porosity is created thanks to this [4]. An increase in the density and fine-grainedness of the metal produced by this technology can be achieved by the dispersive hardening of the crystallizing melt by refractory particles [6, 7]. These methods of increasing the density of metal using known technologies of metallurgical production were taken by the authors as a basis in this work.

2. Carrying out experiments to obtain ingots using the ESR method

The production of solid ingots using the ESR method using classical technology and technology with rotation of the consumable electrode was performed on the upgraded A-550 unit. For remelting, electrodes were used from experimental steels No. 1 and No. 2 with a diameter of 40 mm having a length of 1500 mm, the chemical composition of the remelted steels is shown in table 1.

Table 1. Chemical composition of experimental steel grades (wt. %).

	C	P	S	Mn	Si	Ni	Cr	Cu	Mo	W	Ti
No. 1	0.195	0.025	0.003	0.449	0.802	0.247	14.107	0.059	0.022	0.038	0.057
No. 2	0.169	0.031	0.005	0.306	0.624	5.497	33.103	0.086	0.142	0.082	0.389

The electrodes were attached to the movable carriage of the electroslag remelting installation using a collet clamp. After fixing the remelted electrode, current-carrying brushes were pressed against it using springs to provide a larger contact surface. The remelting was carried out in a crystallizer with a diameter of 90 mm.

The flux of ANF-6, whose chemical composition is presented in table 2, was filled in the working space of the water-cooled crystallizer before the start. The flux granules were previously sieved and had a fraction of not more than 3 mm; also the flux was pre-calcined for two hours at 400 °C to remove residual moisture from it.

Table 2. Chemical composition of the ANF-6 flux (wt. %).

CaF_2	Al_2O_3	CaO	SiO_2	C	TiO_2	Fe_2O_3	S	P
Basis	25	No more						
	31	8	2.5	0.10	0.05	0.5	0.05	0.02

The melting period lasted until the electrode was completely melted down and was 30 minutes. The melted end of the remelted electrode in case of using the classical technology had a conical shape, characteristic for the ESR process. With the use of technology with rotation of the consumable electrode, the beginning of the melting was carried out in a similar way, the rotation started from the moment of formation of the liquid bath. This was evidenced by the absence of current jumps. The rotation speed during the experiments was 100 rpm. Received ingots were given the following marking:

1 - ingot from the experimental grade of steel No. 1, the classical technology of remelting;
2 - ingot from the experimental grade of steel No. 2, the classical technology of remelting;
3 - ingot from the experimental grade of steel No. 1, technology of remelting with rotation of the consumable electrode;
4 - ingot from the experimental grade of steel No. 2, technology of remelting with rotation of the consumable electrode.

Before the extraction of all the ingots obtained, the slag was allowed to cool, and the time for its complete solidification was 8 minutes. After the slag solidified, the ingots were removed from the crystallizer. In all cases ingots had a satisfactory surface quality. Parameters of the obtained ingots of ESR according to different technologies are presented in table 3.

Table 3. Parameters of received ingots of ESR according to different technologies.

	Diameter (mm)	Height (mm)	Weight (kg)
1	90	225	11.27
2		210	10.72
3		230	11.28
4		180	10.47

A study was made of the structure of experimental materials with a microscope for the study of Axio Observer.D1m materials using an image analysis system for solving metallurgical problems and controlling the quality of Thixomet [8]. Examination of samples from experimental ingots of the ESR showed that there are differences in the structure of the ingot microstructures obtained by the classical technology and ingots obtained using the rotation of the consumable electrode. When comparing the structure of the ingot structure from the same experimental grade of steel smelted by different methods (classical technology and rotation of the electrode), it is observed that technology with the use of electrode rotation makes it possible to obtain a structure more uniform and shallower than using classical technology 'figure 1'.

There are also differences in the structure of the ingot structure from the experimental steel grade No. 1 and No. 2: the structure of the ingot from the experimental steel grade No. 2 when using both melting technologies has a more uniform and fine structure compared to the ingots from the experimental grade No. 2 steel.

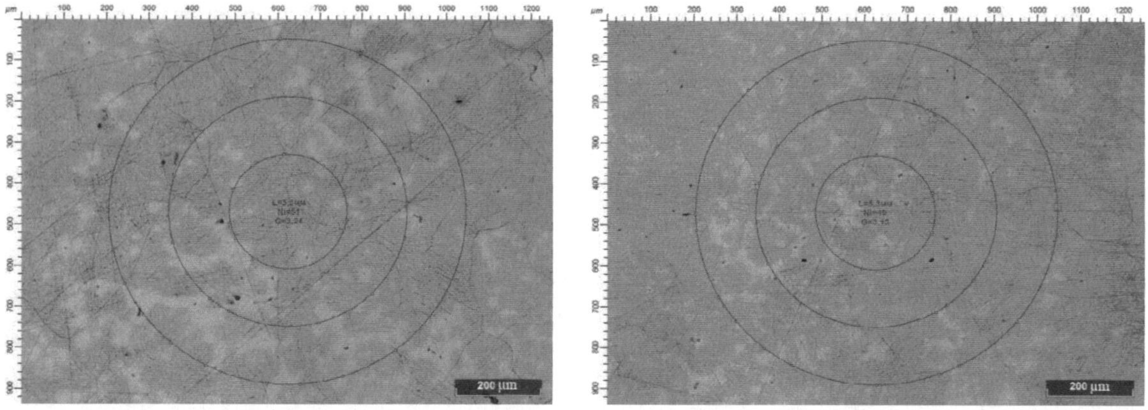

(a) (b)

Figure 1. Microstructure of samples obtained by different ESR technologies, ×130: (a) - ingot No. 1, (b) - ingot No. 3.

3. Carrying out experiments on obtaining ingots by the method of centrifugal casting

The experimental steel grade No. 1, whose chemical composition is presented in table 3, was used as the charge materials for experiments on the production of experimental materials. The charge materials were selected in a certain form to ensure maximum filling of the crucible and, as a consequence, to ensure its maximum productivity. The charge was pre-dried at a temperature of 200 °C before melting.

Melting of charge materials for the production of cylindrical ingots was carried out in an induction melting furnace SELT-001-40/12-T. The time for complete melting and bringing the liquid melt to the required temperature was 25 minutes. WC and TiC powders having a dimension of 2-4 µ m, a density of 15.8 g/cm^3 and 3.21 g/cm^3, respectively, were selected as reinforcing dispersed particles. The crystallization time of the melt in the mould was to be increased in order to ensure an even distribution of the particles over the entire section of the billet, for this, the mould and the poured gutter were preheated to a temperature of 400 °C, and the temperature of the metal to be filled was 1650 °C. The launch of a horizontal centrifugal casting plant was carried out after the metal was brought to the optimum discharge temperature and slag removal. The rotational speed of the mould for casting was 700 rpm. The supply of dispersed particles was carried out directly into the mould by a pneumatic metering device. The weight of the metal poured into the mould was 6 kg per ingot.

Five castings were obtained with different concentrations of introduced particles during the run-up of cylindrical ingots from the experimental grade of steel No. 1 with the introduction of dispersed hardening particles. The received markings were given the following marking:

5 - casting without the introduction of dispersed particles (reference sample);
6 - casting with the introduction of WC, concentration - 1 wt. %;
7 - casting with the introduction of TiC, concentration - 1 wt. %;
8 - casting with the introduction of WC, concentration - 2 wt. %;
9 - casting with the introduction of TiC, concentration - 2 wt. %.

The blanks were placed in quartz sand for uniform cooling after the extraction of all the obtained centrifugally cast blanks from the ingot mould. The parameters of the ingots obtained are presented in table 4.

Table 4. Parameters of ingots obtained by the method of centrifugal casting.

	Height of ingot (mm)	Outer diameter (mm)	Inside diameter (mm)	Weight (kg)
5			110	5.69
6			116	5.78
7	65	140	104	5.68
8			106	5.52
9			112	5.65

Examination of samples from ingots obtained by the method of centrifugal casting showed that the introduction of dispersed hardening particles grinds the structure of the metal, making it more uniform, since refractory particles are not soluble in the metal; they are additional centres of crystallization, thereby increasing the rate of solidification of the metal 'figure 2'. The increase in the grain score does not depend on their composition with an increase in the concentration of introduced dispersed particles.

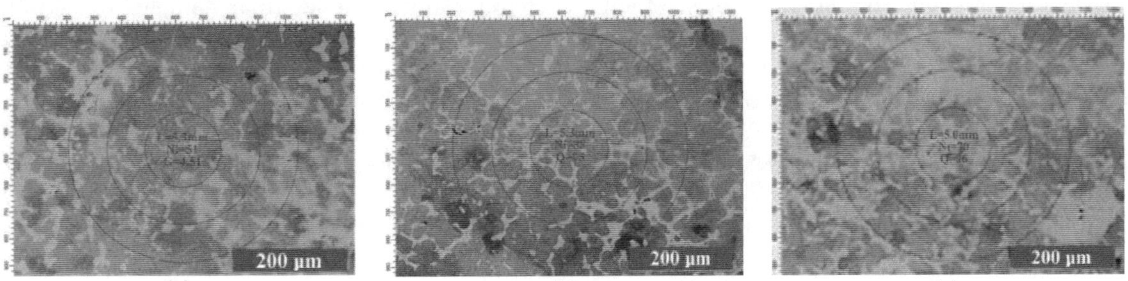

(a) (b) (c)

Figure 2. Microstructure of samples obtained by centrifugal casting, ×130: (a) - ingot No.5, (b) - ingot No.6, (c) - ingot No.8.

4. Carrying out works to assess the corrosion resistance of the samples obtained

Samples 4 mm in height, 10 mm in thickness and 40 mm in length were made to conduct work to assess the corrosion resistance of the obtained samples from each ingot using the HONOR SEIKI VL-100CR lathe and carousel. Estimation of corrosion resistance was carried out with the help of the autoclaved test complex "Cortest" in corrosive-aggressive conditions:
- mineralization - 50 g/l;
- pH - 2.5...3;
- partial pressure of CO2 - 3.0 MPa;
- the total pressure - 5.0 MPa;
- test temperature - 120 °C;
- concentration of CO_2 at the beginning / end of the test - 0,086 / 0,055 g/l;

- duration of the test - 120 hours.

Determination of the rate of general corrosion was carried out by gravimetric method in accordance with GOST 9.908 by determining the mass loss of a sample per unit surface during its exposure to an aggressive medium. The data obtained are presented in table 5.

Table 5. Rate of total corrosion of samples from experimental ingots.

Sample marking	Corrosion rate, mm/year
1	0.24
2	0.00
3	0.01
4	0.09
5	0.73
6	0.70
7	0.32
8	1.46
9	0.33

5. Conclusion

The conclusion can be made on the basis of the results obtained, that the most promising method for increasing the metal density in order to increase the corrosion resistance is the ESR technology, while comparing the corrosion rates of the experimental steels Nos. 1 and 2, the data diverge. The experimental grade of steel No. 1 has the best corrosion resistance index when using the ESR technology with the use of the rotated electrode rotation (the corrosion rate of sample No. 3 (0.01 mm/year) is less than sample No. 1 (0.24 mm/year), and the experimental grade steel No. 2 - using classical technology (corrosion rate of sample No. 2 (0.00 mm/year) is less than sample No. 4 (0.09 mm/year).

Cylindrical samples obtained by the method of centrifugal casting with the introduction of hardening particles (samples No. 5-9) have corrosion resistance values lower than ESR ingots. Samples modified with WC tungsten carbide have corrosion resistance values almost equal to the standard (standard 0.73 mm/year, 1% WC - 0.70 mm/year) or less (2% WC - 1.46 mm/year). When the metal is modified with TiC dispersible particles, the corrosion resistance increases: the comparison sample has a value of 0.73 mm/year; the modified samples have 0.32 mm/year (1% TiC) and 0.33 mm/year (2% TiC). It is noteworthy that an increase in the concentration of TiC particles to 2% does not give an additional effect of increasing corrosion resistance.

Thus, according to the work done, it can be concluded that the use of titanium carbide as a modifier in minor amounts allows a more than 2-fold increase in the resistance to corrosion. All experimental samples (except modified with tungsten carbide) have values corresponding to the accepted norm for a number of parts used in the oil and gas industry - 0.5 mm/year.

Acknowledgments

South Ural State University is grateful for financial support of the Ministry of Education and Science of the Russian Federation (grant No 11.9658.2017/8.9.)

References

[1] Khaustov A P and Redina M M 2010 *Law and security* **2** 86
[2] Duckworth U E and Hoyle J 1973 *Electroslag remelting* (Moscow: Metallurgy)
[3] Pribulová A, Futáš P, Kmita A, Márosová D and Holtzer M 2017 *Archives of Metallurgy and Materials* **62** 181

IOP Conf. Series: Materials Science and Engineering **265** (2017) 012033 doi:10.1088/1757-899X/265/1/012033

[4] Yudin S B, Levin M M and Rosenfeld S E 1972 *Centrifugal casting* (Moskow: Mechanical Engineering)
[5] Chumanov I V and Chumanov V I 2001 *Metallurgist* **45** 125
[6] Chumanov V I, Chumanov I V, Anikeev A N and Garifulin R R 2010 *Russian Metallurgy (Metally)* **2010** 1125
[7] Sijo M T, Jayadevan K R and Janardhanan S 2017 *International Journal of Mechanical Engineering and Technology* **8** 66
[8] Internet-resource: http://www.thixomet.ru/products/?show=2 (date of the application 29.07.2017)

Analysis of SCC sensitive areas for X65 pipeline structure in marine environment

L Y Bai[1], H Zhou[2], L Gao[1], Q Yin[1] and K B Jiang[1*]

[1] Field Engineering Institute of PLA University of Science and Technology, Nanjing Jiangsu, 210007, China

[2] Wuhan Ship Development Design Institute, Wuhan Hubei, 430000, China

kbjiang1122@sina.com

Abstract. SCC sensitive areas of welded specimens in the marine environment was investigated by means of magnetic field distribution detection, residual stress measurement and corrosion morphology observation. The study found that both HAZ and parent material surfaces of welded specimens could generate high-level stress concentration areas under the action of marine environment, thereby becoming sensitive areas of SCC. The combined effect of residual stress and corrosion damage results in the formation of SCC sensitive areas. Meanwhile, SCC sensibility on HAZ and parent material surfaces are determined by residual stress level and corrosion damage degree on the structure surface, respectively. Appropriate preventive measures should be taken according to the determinant of HAZ and parent material surfaces' SCC sensibility when SCC preventive measures of X65 pipeline structure servicing in the marine environment are designed.

1. Introduction

Pipeline transportation, which has been a main transportation mode of oil and gas, was usually damaged by stress corrosion cracking (SCC) while servicing in the marine environment, reducing the security and reliability of pipeline structure.[1-3] A number of studies indicated that areas on the heat affected zone (HAZ) surface had high-incidence of SCC, and the problems about influence of residual stress on the stress corrosion behavior of welded structure has become a research emphasis in recent years, consequently.[4-6] According to Ravi Kiran,[7] the damage fracture of ductile metal is developed from nucleation, growth, and convergence. Hence, the corrosive microvoid generating on the parent metal surface could also develop into corrosion damage while servicing in the marine environment, resulting in the SCC and influence the safe use of pipeline structure which is mainly jointed by welding.

Microvoids which is in nucleation stage is difficult to be probed by instruments. However, magnetic field leakage will be produced because of stress concentration generating in the growth and convergence stages of microvoids, thus magnetic leakage detection becomes an effective method to probe earlier corrosion manage of ductile metal. Stress concentration areas on the X65 welded specimens' surfaces were measured after corrosion in the marine environment, and the forming reasons of stress concentration areas generating on HAZ and parent metal surfaces were analyzed through residual stress measurement and corrosion morphology observation. The SCC sensitive areas of X65 pipeline structure in marine environment were studied, the results of which have great reference significance on design of SCC preventive measures of X65 pipeline structure servicing in marine environment.

2. Experimental

The chemical components of X65 used in the experiments is 0.082 C-0.009 S-0.024 P-0.21 Si-1.42 Mn, and the size of welded specimens is 350mm×150mm×10mm. The corrosion experiment with three welded specimens whose number is A1-A3 was conducted in accordance with GB/T 6384-2008 in natural seawater for 60 days. To increase the corrosion rates of welded specimens in seawater, the specimens were connected to graphite sheets. After corrosion experiment, the corrosion products were removed by acid washing, and an inhibitor was added in the acid solution to prevent reactions between corroded specimens and acid solution.

Magnetic determinator, whose type is TSC-2M-8, was used to measure the magnetic field on corrode specimens' surfaces, aiming to ensure stress concentration areas on the corrode specimens' surfaces through characteristic curve of leakage magnetic field. The compensation of magnetic determinator was opened while measuring, and the parameters of scanning step and measurement distance were 2mm and 4mm, respectively.

Residual stress distribution on corrode specimens' surfaces were measured by hole-drilling technique. In order to paste strain gauges effectively, the surfaces of corrode specimens need to be polished through high speed spindle. Hence, the additional deformation caused by polishing should be measured, using for correct the residual stress measurements of corrode specimens. The aperture and hole-depth used in the drilling process were 2mm and 1.5mm, respectively.

Corrosion morphologies of corrode specimens were observed by a type SNE 4500M scanning electron microscopy (SEM), aiming to ensure the corrosion morphologies' characteristics of stress concentration areas on the HAZ and parent metal surfaces. Thereby, the forming reasons of stress concentration areas on the corrode specimens' surfaces were achieved according to the results of residual stress measurement.

3. Results and discussion

3.1. Magnetic field measurements

Magnetic field distribution of corrode specimens' central regions were detected, aiming to ensure the stress concentration areas on the corrode specimens' surfaces. The magnetic field and magnetic field gradient are shown in figure 1. Where, Hp-1, Hp-3, Hp-5 and Hp-7 represent tangential component of leakage magnetic field (H_p^x); Hp-2, Hp-4, Hp-6 and Hp-8 represent normal component of leakage magnetic field (H_p^y).

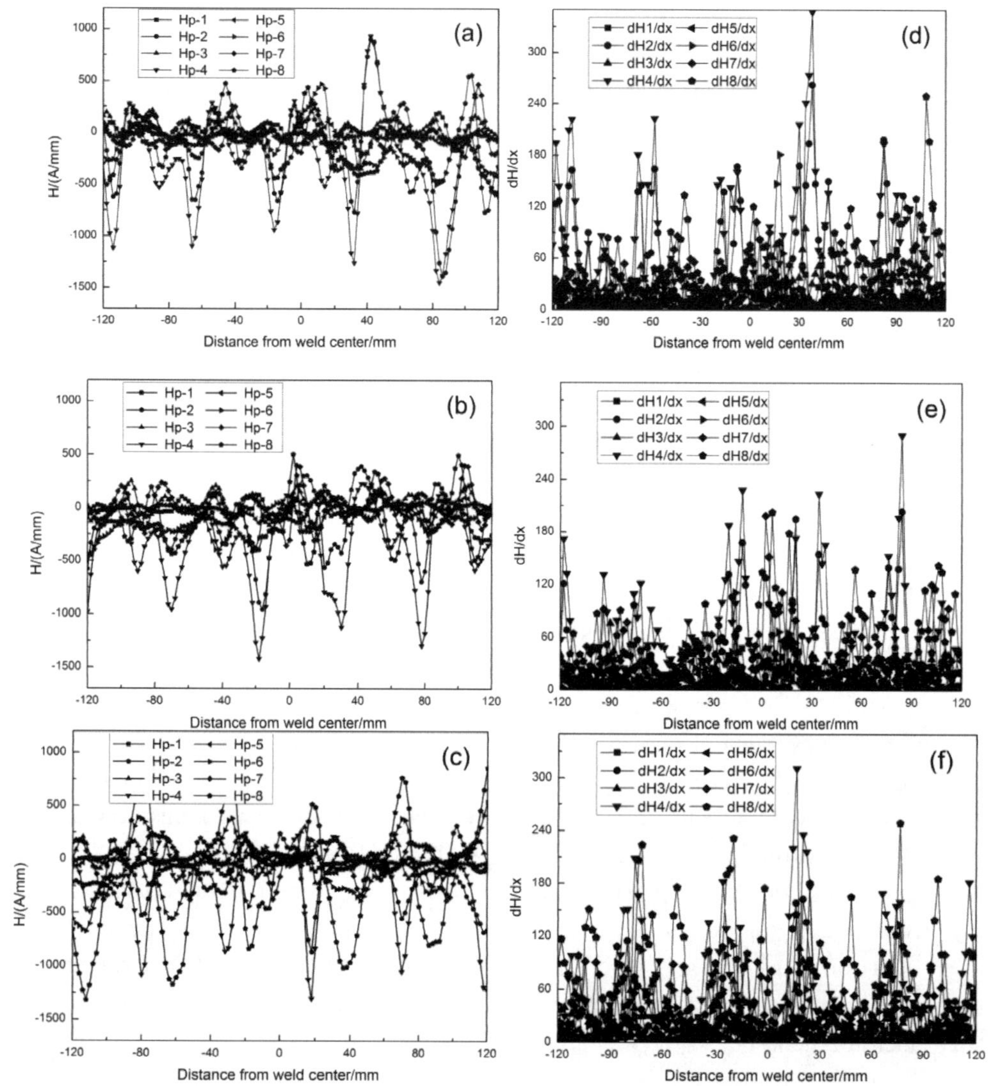

Figure 1. Magnetic field measurements: (a) magnetic field of A1, (b) magnetic field of A2, (c) magnetic field of A3, (d) magnetic field gradient of A1, (e) magnetic field gradient of A2, (f) magnetic field gradient of A3.

As shown in figure 1(a) - figure 1(c), the fluctuation exists in multiple parts of H_p^x and H_p^y on corrode specimens' surfaces, and the characteristic form of leakage magnetic field, where H_p^x reaches a maximum and H_p^y passes through zero, appears in many multiple areas. The characteristic form of leakage magnetic field distributes on both HAZ and parent metal surfaces, which is indicate that numbers of leakage magnetic points are produced on the HAZ and parent metal surfaces during corrosion process.

Magnetic field gradient of multiple leakage magnetic points on the corrode specimens' surfaces presents a higher level as shown in figure 1(d) –figure 1(f). The peaks of magnetic field gradient on A1 and A3 surfaces are located in HAZ, while the peak of magnetic field gradient on A2 surface is located in parent metal. Therefore, high-level stress concentration areas are generated on both HAZ and parent metal surfaces of welded structure, because of the level of magnetic field gradient is proportional to the level of stress concentration.

To observe the distribution characteristics of stress concentration areas on corrode specimens' surfaces, Metal Magnetic Method System 3.0 was used to process the data of magnetic field gradient on the corrode specimens' surfaces, as shown in figure 2.

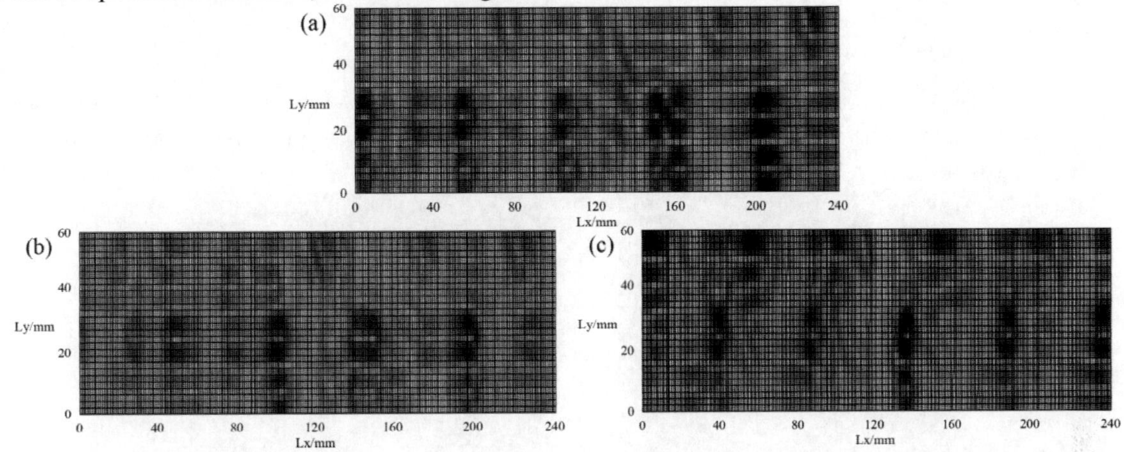

Figure 2. Magnetic field gradient distributions of corrode specimens' surfaces: (a) magnetic field gradient distribution of A1 surface, (b) magnetic field gradient distribution of A2 surface, (c) magnetic field gradient distribution of A3 surface.

In figure 2, the center of welded joint is the position Lx=120mm, and the dark areas represent the areas with higher stress concentration level, while the light areas represent the areas with lower stress concentration level.

The corrode specimens of A1-A3 are totally have high-level stress concentration areas on HAZ surfaces, as shown in figure 2. However, the position and degree of stress concentration areas on the parent metal surfaces of A1-A3 are irregular, stress concentration level on part of areas on the parent metal surfaces of A1-A3 are even larger than that on HAZ surfaces.

3.2. Residual stress measurements

The residual stress measurements of corrode specimens are shown in figure 3.

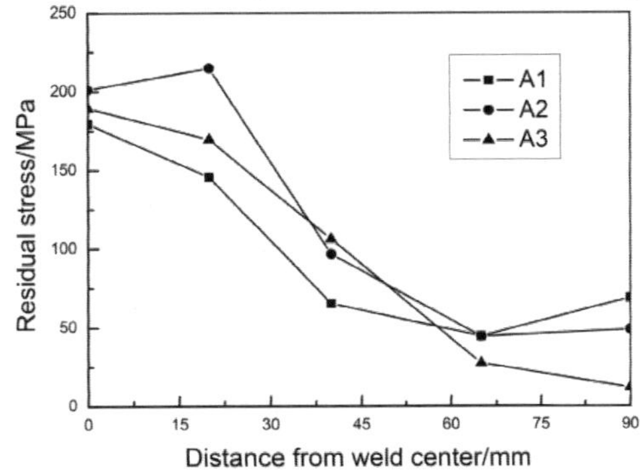

Figure 3. Residual stress measurements

As shown in figure 3, the residual stress level on the HAZ surfaces of corrode specimens is higher than other areas, and the residual stress peaks of A1-A3 are 179.5MPa, 214.7MPa and 189.4MPa, respectively。Residual stress on the corrode specimens' surfaces decrease with increase of the distance from weld center, then remain a low level on the parent metal surface.

Hence, the generation of high-level stress concentration areas on the HAZ surface of corrode specimen are related to residual stress, while the residual stress level on the parent metal surface is much lower than that on the HAZ surface. Relying only on the residual stress on the parent metal surface is not enough to arouse the generation of high-level stress concentration areas.

3.3. Corrosion morphology measurements

A1 was selected as specimen for corrosion morphology observation to research the corrosion morphologies' characteristics on the HAZ and parent metal surfaces of corrode specimen, the corrosion morphology of A1 is shown in figure 4.

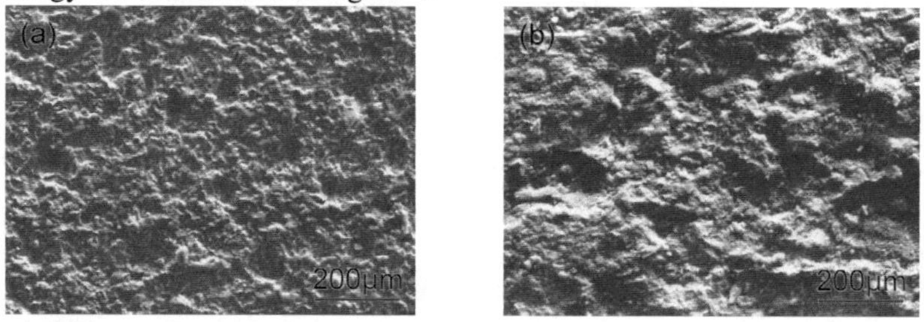

Figure 4. Surface corrosion morphologies in different areas of A1: (a) HAZ surface, (b) parent metal surface

As shown in figure 4, the corrosion morphology on HAZ surface resembles uniform corrosion, which is accompanied by a small amount of etch pits. Most of pits are in the growth stage of microvoids, whose diameters are in the range of 100μm to 150μm. Small part of pits converge together, entering into the convergence stage of microvoids. Thus, although corrosion damage on the HAZ surface have stimulative effect to the generation of stress concentration areas, the impact of corrosion manage on the generation of stress concentration areas are much lower than that of residual stress. Corrosion morphology on the parent metal surface is based on pitting and the etch pits with large diameter appears intensively in many areas, severe corrosion damage ultimately forms at the confluence of etch pits. Hence, the generation of stress concentration areas on the parent metal surface is mainly caused by corrosion damage rather than low-level residual stress.

In combination with the measurements of residual stress we can realize that the stress concentration areas generating on the HAZ and parent metal surfaces of corrode specimen is owing to the combined influence of residual stress and corrosion manage. Stress concentration areas generating on the HAZ surface are mainly caused by residual stress, while stress concentration areas generating on the parent metal surface are mainly caused by corrosion manage. Hence, appropriate preventive measures should be taken according to the determinant of HAZ and parent material surfaces' SCC sensibility when SCC preventive measures of X65 welded pipeline structure servicing under the marine environment are designed.

4. Conclusion

1. High-level stress concentration areas are produced on both HAZ and parent metal surfaces of welded structure during corrosion, increasing the SCC sensibility of welded structure. The SCC failure could occur on both HAZ and parent metal of X65 pipeline structure servicing in the marine environment under the action of corrosion.

2. The production of SCC sensitive areas is the results of combined effect of residual stress and corrosion manage. SCC sensitive areas generating in HAZ is determined by residual stress, while SCC sensitive areas generating in parent metal is determined by corrosion damage.
3. Appropriate preventive measures should be taken according to the determinant of HAZ and parent material surfaces' SCC sensibility when SCC preventive measures of X65 pipeline structure servicing in the marine environment are designed.

Acknowledgements

This research work was financial supported by the Major State Basic Research Development of China (973 Program, Grant Nos. 2014CB046801).

Reference

[1] Hertelé S, Cosham A and Roovers P 2016 *Eng. Struct.* **124** 429–41
[2] Nakamura N, Ashida K, Takishita T, Ogi H and Hirao M 2016 *NDT&E Int.* **83** 88–93
[3] Fan Z, Liu J Y, Li S L and Zhang T J 2009 *J. Southwest Petrol. Univ.* **31(5)** 171–74.
[4] Ma H C, Liu Z Y, Du C W, Wang H R, Li X G, Zhang D W and Cui Z Y 2015 *Corros. Sci.* **8** 6467–82
[5] Lim Y S, Hwang S S, Kim S W and Kim H P 2015 *Corros. Sci.* **6** 2–45
[6] Lu Q, Gu B S, Yang P Y and Wang Y 2015 *Corros. Prot.* **36(10)** 1004–08
[7] Kiran R and Khandelwal K 2015 *Eng. Fract. Mech.* **6** 1–70